Mastering the Arduino Uno R4

Programming and Projects for the Minima and WiFi

Dogan Ibrahim

elektor
design > share > earn

This is an Elektor Publication. Elektor is the media brand of
Elektor International Media B.V.
PO Box 11, NL-6114-ZG Susteren, The Netherlands
Phone: +31 46 4389444

Declaration

British Library Cataloguing in Publication Data
A catalogue record for this book is available from the British Library

ISBN 978-3-89576-578-0 Print
ISBN 978-3-89576-579-7 eBook

Editors: Alina Neacsu; Jan Buiting MA
Prepress Production: D-Vision, Julian van den Berg
Print: Ipskamp Printing, Enschede (NL)

Contents

Preface **11**

Chapter 1 • The Arduino Uno R4 . . . . **12**

1.1 Overview 12

1.2 The Arduino Uno R4 against Uno R3 13

1.3 The Arduino Uno R4 Minima hardware 15

1.4 The Arduino Uno R4 Projects Kit 20

Chapter 2 • Arduino Uno R4 Program Development **23**

2.1 Overview 23

2.2 Installing the Arduino IDE 2.1.0 24

2.3 Software-only programs 26

2.3.1 Example 1: Sum of integer numbers 26

2.3.2 Example 2: Table of squares 29

2.3.3 Example 3: Volume of a cylinder 30

2.3.4 Example 4: Centigrade to Fahrenheit 31

2.3.5 Example 5: Times table 33

2.3.6 Example 6: Table of trigonometric sine 34

2.3.7 Example 7: Table of trigonometric sine, cosine and tangent 36

2.3.8 Example 8: Integer calculator 37

2.3.9 Example 9: Dice 40

2.3.10 Example 10: Floating point calculator 41

2.3.11 Example 11: Binary, octal, hexadecimal 43

2.3.12 Example 12: String functions 44

2.3.13 Example 13: Initializing an array 46

2.3.14 Example 14: Character functions 48

2.3.15 Example 15: Solution of a quadratic equation 50

2.3.16 Example 16: Lucky day of the week 53

2.3.17 Example 17: Factorial of a number 54

2.3.18 Example 18: Add two square matrices 56

Chapter 3 • Hardware Projects with LEDs **59**

3.1 Overview 59

3.2 Project 1: Blinking LED – using the on-board LED 59
3.3 Project 2: Blinking LED – using an external LED 60
3.4 Project 3: LED flashing SOS 63
3.5 Project 4: Alternately blinking LEDs 64
3.6 Project 5: Chaser-LEDs 67
3.7 Project 6: Chasing LEDs 2 70
3.8 Project 7: Binary counting LEDs 72
3.9 Project 8: Random flashing LEDs - Christmas lights 74
3.10 Project 9: Button controlled LED 75
3.11 Project 10: Controlling the LED flashing rate - external interrupts 80
3.12 Project 11: Reaction timer 83
3.13 Project 12: LED color wand 85
3.14 Project 13: RGB fixed colors 87
3.15 Project 14: Traffic lights 89
3.16 Project 15: Traffic lights with pedestrian crossings 94
3.17 Project 16: Using the 74HC595 shift register – binary up counter 100
3.18 Project 17: Using the 74HC595 shift register - random flashing 8 LEDs 103
3.19 Project 18: Using the 74HC595 shift register - chasing LEDs 104
3.20 Project 19: Using the 74HC595 shift register - turn ON a specified LED 105
3.21 Project 20: Using the 74HC595 shift register - turn ON specified LEDs 107

Chapter 4 • 7-Segment LED Displays 109

4.1 Overview 109
4.2 7-Segment LED display structure 109
4.3 Project 1: 7-Segment 1-digit LED counter 111
4.4 Project 2: 7-Segment 4-digit multiplexed LED display 114
4.5 Project 3: 7-Segment 4-digit multiplexed LED display counter - timer interrupts 119
4.6 Project 4: 7-Segment 4-digit multiplexed LED display counter - blanking leading zeroes 124
4.7 Project 5: 7-Segment 4-digit multiplexed LED display - reaction timer 128
4.8 Project 6: Timer interrupt blinking on-board LED 133

Chapter 5 • Liquid Crystal Displays 136

5.1 Overview 136

5.2 The I^2C bus. 136
5.3 I^2C ports of the development board 137
5.4 I^2C LCD 137
5.5 Project 1: Display text on the LCD 140
5.6 Project 2: Scrolling text on the LCD 142
5.7 Project 3: Display custom characters on the LCD. 144
5.8 Project 4: LCD based conveyor belt goods counter. 145
5.9 Project 5: LCD based accurate clock using timer interrupts. 149
5.10 Project 6: LCD dice 154

Chapter 6 • Sensors 157

6.1 Overview 157
6.2 Project 1: Analog temperature sensor 157
6.3 Project 2: Voltmeter. 160
6.4 Project 3: On/off temperature controller. 161
6.5 Project 4: Darkness reminder – using a light-dependent resistor (LDR) 164
6.6 Project 5: Tilt detection 167
6.7 Water level sensor. 169
6.7.1 Project 6: Displaying water level. 169
6.7.2 Project 7: Water level controller 172
6.7.3 Project 8: Water flooding detector with buzzer 174
6.8 Project 9: Sound detection sensor — control the relay by clapping hands. 175
6.9 Project 10: Flame sensor - fire detection with relay output 177
6.10 Project 11: Temperature and humidity display. 180
6.11 Project 12: Generating musical tones - melody maker 184

Chapter 7 • The RFID Reader 187

7.1 Overview 187
7.2 Project 1: Finding the Tag ID 187
7.3 Project 2: RFID door lock access with relay. 190

Chapter 8 • The 4×4 Keypad 194

8.1 Overview 194
8.2 Project 1: Display the pressed key code on the Serial Monitor. 195
8.3 Project 2: Integer calculator with LCD 198

8.4 Project 3: Keypad door security lock with relay 203

Chapter 9 • The Real-Time Clock (RTC) Module 207

9.1 Overview 207

9.2 The supplied RTC module 207

9.3 Project 1: RTC with Serial Monitor 207

9.4 Project 2: RTC with LCD 211

9.5 Project 3: Temperature and humidity display with time stamping 213

9.6 Using the built-in RTC 216

9.6.1 Project 4: Setting and displaying the current time 216

9.6.2 Project 5: Periodic interrupt every 2 seconds 218

Chapter 10 • The Joystick. 221

10.1 Overview 221

10.2 The joystick 221

10.3 Project 1 - Reading analog values from the joystick 221

Chapter 11 • The 8×8 LED Matrix. 226

11.1 Overview 226

11.2 The supplied 8×8 LED matrix 226

11.3 Project 1: Displaying shapes 227

Chapter 12 • Motors: Servo and Stepper 231

12.1 Overview 231

12.2 The servo motor 231

12.2.1 Project 1: Test-rotate the servo 232

12.2.2 Project 2: Servo sweep 234

12.2.3 Project 3: Joystick-controlled servo 235

12.3 The stepper motor 237

12.3.1 Project 4: Rotate the motor clockwise and then anticlockwise. 238

Chapter 13 • The Digital To Analog Converter (DAC). 241

13.1 Overview 241

13.2 Project 1: Generating a square wave with 2 V amplitude 241

13.3 Generating sine wave – using the analogWave library 242

13.3.1 Project 2: Generate a sine wave 243

13.3.2 Project 3: Sine wave sweep frequency generator 244

13.3.3 Project 4: Generate sine wave whose frequency changes with potentiometer. . 245

13.3.4 Project 5: Generate a square wave with frequency of 1 kHz and amplitude of 1 V 247

Chapter 14 • Using the EEPROM, the Human Interface Device, and PWM 248

14.1 Overview 248

14.2 The EEPROM memory. 248

14.3 Human Interface Device (HID) 249

14.4 Project 1: Keyboard control to launch Windows programs. 250

14.5 The Pulse Width Modulation (PWM) 253

14.5.1 PWM channels of the Arduino Uno R4 255

14.5.2 Project 2: LED dimming using PWM. 255

Chapter 15 • The Arduino Uno R4 WiFi 257

15.1 Overview 257

15.2 The LED matrix 260

15.2.1 Project 1: Using LED matrix 1 - creating a large + shape. 260

15.2.2 Project 2: Creating images by setting bits 262

15.2.3 Project 3: Using LED matrix 2 - creating a large + shape. 265

15.2.4 Project 4: Animation - displaying a word 267

15.3 Using the WiFi. 269

15.3.1 UDP and TCP 269

15.3.2 UDP communication 270

15.3.3 TCP communication. 271

15.3.4 Project 5: Controlling the Arduino Uno R4 WiFi on-board LED from a smartphone using UDP 272

15.4 Bluetooth 276

15.4.1 Bluetooth BLE. 277

15.4.2 Bluetooth BLE Software Model 278

Chapter 16 • Serial Communications 280

16.1 Overview 280

16.2 Project 1: Receiving ambient temperature from an Arduino Uno R3 281

Chapter 17 • Using an Arduino Uno Simulator. 285

17.1 Why simulation? 285
17.2 The Wokwi simulator 286
17.2.1 Project 1: A simple project simulation - flashing LED. 287
17.2.2 Project 2: Displaying text on LCD 288
17.2.3 Project 3: LCD seconds counter 290
Chapter 18 • The CAN bus 292
18.1 Overview 292
18.2 The CAN bus 292
18.2.1 CAN bus termination 292
18.2.2 CAN bus data rate. 294
18.2.3 Cable stub length 295
18.2.4 CAN bus node. 295
18.2.5 CAN bus signal levels 296
18.2.6 CAN_H voltage 297
18.2.7 The CAN_L voltage 297
18.2.8 Bus arbitration 297
18.2.9 Bus transceiver. 297
18.2.10 CAN connectors 298
18.3 Arduino Uno R4 CAN bus interface 300
18.3.1 CAN bus transceivers. 300
18.4 Project 1: Arduino Uno R4 WiFi to Arduino Uno R4 Minima CAN bus communication 301
18.5 Project 2: Sending the temperature readings over the CAN bus. 306
Chapter 19 • Infrared Receiver and Remote Controller 311
19.1 Overview 311
19.2 The supplied infrared receiver 311
19.3 The supplied infrared remote control transmitter unit. 311
19.4 Operation of an infrared remote control system. 312
19.5 Project 1: Decoding the IR remote control codes 314
19.6 Project 2: Remote relay activation/deactivation. 317
19.7 Project 3: Infrared remote stepper motor control. 320
Index 325

Preface

"Arduino" is an open-source microcontroller development system that incorporates system hardware, an Integrated Development Environment (IDE), and a large number of libraries. Arduino is supported by a large community of programmers, electronic engineers, enthusiasts, and academics. There are several distinctive designs of the basic Arduino board. Although they are intended for diverse types of applications, they can all be programmed using the same IDE, and, in general, programs can be transported ("ported") between different boards. This is probably one of the reasons for the popularity of the Arduino family. Arduino is also supported by a large number of software libraries for many interface devices that can easily be included in your programs. Using these libraries makes programming relatively easy and speeds up the programming time. Using libraries also makes it easier to test your programs since most libraries have already been fully tested and working.

The **Arduino Uno R3** board probably ranks as the most popular Arduino family member to date and has been with us for many years. Based on the low-cost 8-bit ATmega328P processor, the Uno R3 has been used by students and hobbyists in many beginning and intermediate-level, low-speed projects requiring small to medium amounts of memory. Perhaps one of the attractive points of the Uno R3 was its powerful IDE and the simplicity of using it to develop projects in relatively short times.

Recently, the new Arduino Uno R4 was announced. This new board is compatible with the earlier Uno R3 but offers highly improved specifications compared to Uno R3. The new **Arduino Uno R4** is based on a 48 MHz 32-bit Cortex-M4 processor with a large amount of SRAM and flash memory. Additionally, a higher-precision ADC and a new DAC are added to the design. The contemporary design also supports the CAN bus interface. Two versions of the board are available: **Uno R4 Minima**, and **Uno R4 WiFi**.

This book is about using these new boards to develop many different and interesting projects. The projects given in the book have been fully tested with just a handful of parts and external modules, which are available as a kit from Elektor. The block diagrams, circuit diagrams, full program listings, and complete program descriptions are given for all the projects in the book. You should find it easy to build the project hardware and then follow the software descriptions given for the projects. It should also be relatively easy to modify the hardware and software for your own project applications.

I hope that you enjoy reading the book and at the same time learn how to use the Arduino Uno R4 Minima or the R4 WiFi models in your innovative projects.

Dogan Ibrahim
London, 2023

Publisher's Notice: All programs discussed in this Guide are contained in an archive file you can download free of charge from the Elektor website. Head to www.elektor.com/books and enter the book title in the Search box.

Chapter 1 • The Arduino Uno R4

1.1 Overview

The Arduino project started out as a tool for students at the **Interactive Design Institute**, **Ivrea** in Italy back in 2005. The aim of this project was to provide low-cost and easy-to-use hardware and software to beginner students and hobbyists to create simple projects using sensors and actuators. The name "Arduino" comes from the bar named *Arduin of Ivrea* where the project's founders used to meet for drinks. The name of this bar came from the Margrave *Arduin of Ivrea*, who was King of Italy from 1002 to 1014.

The initial Arduino project team consisted of Massimo Banzi, David Cuartielles, Tom Igoe, Gianluca Martino, and David Mellis. In 2003 Hernando Barragan created the development platform *Wiring* as a master's thesis at the Institute Ivrea. Wiring was an open-source electronics platform consisting of a programming language, an integrated development environment (IDE), and a single-board microcontroller. This project was developed under the supervision of Massimo Banzi and Casey Reas. The Wiring platform included a printed circuit board with an ATmega128 microcontroller, an IDE, and some library functions. Later in 2005, Massimo Banzi, David Mellis, and David Cuartielles extended the Wiring platform by adding support for the ATmega8 microcontroller which was cheaper. This new project was named Arduino. The project was so successful that after developing less expensive versions, by mid-2011 over 300,000 copies of Arduino were produced commercially and by 2013 this number increased to over 800,000.

Arduino is an open-source hardware where the designs are distributed under a Creative Commons license and are freely available. The only point is that the developers are requested the name Arduino to be reserved for the official product and not be used for similar copy work without permission.

One of the nice features of the Arduino series is that a pre-programmed boot loader is used on the on-board processor. Users can develop their programs using the IDE and then upload their programs to the Arduino processor with the help of this boot loader program. The I/O pins are available at female headers located on either side of the board. This makes the hardware development very easy as jumper wires can be used to make connections to the board.

The original Arduino board was manufactured by the Italian company called Smart Projects. Many versions of the Arduino have been developed over the years by several companies. Some versions are:

- Arduino Diecimila
- Arduino Uno R2
- Arduino Leonardo
- Arduino RS232
- Arduino Pro
- Arduino Mega
- Arduino LilyPad

- Arduino Robot
- Arduino Esplora
- Arduino Yun
- Arduino Fio
- Arduino Ethernet
- Arduino Due
- Arduino Nano
- Arduino Uno SMD R3
- Arduino Uno R3
- Arduino MKR Zero
- Arduino Zero
- ... and many more

The Arduino family has been so popular that in 2022, its revenue amounted to over US$237 million, including a large portion of online sales via the Internet.

Two new versions of the Arduino have recently been announced: **Arduino Uno R4 Minima**, and **Arduino Uno R4 WiFi**. In this book, you will be developing projects using both the Arduino Uno R4 Minima and the Arduino Uno R4 WiFi. The new two boards are similar to the very popular Arduino Uno R3 board but they have been expanded in many ways.

1.2 The Arduino Uno R4 against Uno R3

A comparison of the Uno R3 and Uno R4 is given in Table 1.1. Notice that almost all the projects and libraries used with the Uno R3 can be used with the Uno R4 without any modifications. It is recommended however to upgrade any libraries which may have been modified specifically for the Uno R4.

Feature	Arduino Uno R3	Arduino Uno R4
Processor	ATmega328P	Renesas RA4M1
Word length	8	32
Clock speed	16 MHz	48 MHz
SRAM	2 KB	32 KB
Flash memory	32 KB	256 KB
EEPROM	1 KB	8 KB
Operating voltage	5 V	5 V
Timers	3	10
Capacitive touch sensing	None	27 channels
Temperature sensor	None	1
USB connector	Type B	USB-C
ADC	10-bit	14-bit
DAC	None	12-bit
SPI	1	1

I²C	1	2
Qwiic I²C	None	1 (WiFi version only)
Operating voltage	5 V	5 V
Wi-Fi	None	WiFi version only
RTC	None	1 (WiFi version only)
Human Interface Device	None	Yes
SWD debug	None	1
Bluetooth BLE	None	WiFi version only
CAN bus support	None	1
Op Amp	None	1
128x8 LED matrix	None	WiFi version only
Input voltage	7–12 V	6–24 V
Analog inputs	6	6
PWM pins	6	6
USB	USB-B	USB-C

Table 1.1: Comparison of the Arduino Uno R3 and Uno R4.

The Arduino Uno R4 features the Renesas RA4M1 processor, which is an Arm 32-bit Cortex-M4 processor, running at 48 MHz. The Uno R3 had an ATmega328P processor with only a 16 MHz clock. This is a 3 times increase of the clock speed over the Uno R3. Additionally, Uno R4 has 32 KB SRAM memory compared to 2 KB on the Uno R3. The flash memory of the R4 is 256 KB, compared to only 32 KB on the Uno R3. As a result, more complex projects requiring more memory can be developed with the Uno R4. The USB port on the Uno R3 has been replaced with the currently standard USB-C and the maximum power supply voltage has been increased to 24 V with improved thermal design. The processor operating voltage is still 5 V. Arduino Uno R4 provides a CAN bus interface, allowing devices to be connected and programmed on a CAN bus environment. The ADC converter capacity has been increased from 12 bits to 14 bits on the Uno R4. SPI and I²C bus interfaces have been increased from 1 to 2. The Uno R4 supports Human Interface Device (HID) which enables users to simulate a mouse or keyboard when connected to a computer via a USB cable. Users can send mouse movements or keystrokes to a computer. Additionally, Uno R4 includes a true 12-bit DAC converter. The analogWave library was added to make using the DAC easy. Generating a sine, saw or square wave is as easy as calling a library function. Of course, you can do much more with it. The Uno R4 PCB is hardware compatible with the Uno R3. The pinout, voltage, and form factor are unchanged so that the Uno R4 can easily replace designs that use the Uno R3. The software IDE is also compatible between the Uno R3 and Uno R4, where an effort was made to maximize backward compatibility of the Arduino libraries so that users can use the existing libraries without any modifications. Some libraries which depend heavily on the AVR architecture may need to be re-loaded into the IDE. A public list of such libraries will be provided by Arduino.

Compared to other Cortex-based boards such as the Raspberry Pi Pico, which uses the Cortex M0+, the Cortex-M4 performance is about 6 times better (just to remind you, the Raspberry Pi Pico clock runs at 125 MHz by default). As a result, the Arduino Uno R4 can be used in highly complex real-time projects (e.g. digital control, DSP, AI, etc.) requiring large memory and fast throughput.

As mentioned earlier, Arduino Uno R4 is available in two versions: **WiFi** and **Minima**. The WiFi version is equipped with an Espressif S3 Wi-Fi module, making the board ideal in IoT-based network and Bluetooth BLE applications, as well as in applications requiring Internet connectivity. Additionally, a 128×8 LED matrix is included on the board. The Minima version offers a cost-effective option with no WiFi or Bluetooth connectivity. **In this book, all the projects compile and run on both versions, except that the WiFi and LED Matrix projects only compile and run on the R4 WiFi version.**

In summary, the Uno R4 is a giant leap forward for Arduino as it is a truly remarkable board that will take your microcontroller project experience to the next level. Perhaps one disadvantage of the Uno R4 compared to Uno R3 is its increased price.

1.3 The Arduino Uno R4 Minima hardware

Figure 1.1 shows the Arduino Uno R4 PCB layout. The PCB footpath and the placement of the headers on the PCB are the same as the Uno R3. In the center of the PCB is the RA4M1 microcontroller. On the left-hand side of the PCB, you can see the Reset button, USB-C connector for connecting to a PC, and the barrel connector for supplying external power. At the rear of the PCB, you can see the 6-pin ICSP pins (SPI) and the 10-pin SWD/JTAG pins. An on-board LED is connected to port 13 as in Uno R3. Additionally, yellow LEDs are connected to the serial TX and RX pins to indicate serial data transmission, a green LED indicates power to the board, and a yellow LED indicates the state of the SCK pin.

Figure 1.1: Arduino Uno R4 Minima PCB layout.

On the two sides of the PCB, you have the header connectors as in the Uno R3. The headers have the following pins (see Figure 1.2):

Short header connector:

- 6× 14-bit analog input pins (A0–A5)
- 1× 12-bit DAC (A0)
- 1× Opamp+ (A1), Opamp– (A2), Opamp Out (A3)
- I^2C SDA (A4), I^2C SCL (A5)
- Vin
- GND
- +5 V
- +3.3 V output (output from the RA4M1 VCC_USB pin)
- RESET
- IOREF (reference for the digital logic V. Connected to + 5 V)
- BOOT (mode selection)

Note: Analog pins A0–A5 can also be used as digital pins.

Long header connector:

- 14 × digital pins (D0–D13)
- External interrupt (IRQ00: D2, IRQ01: D3)
- 6 × PWM pins (D3, D5, D6, D9, D10, D11)
- UART (RX:D0, TX:D1)
- SPI (same as on ICSP header. D13: SCK, MISO (CIPO): D12, MOSI (COPI): D11, CS: D10)
- CAN (RX:D5, TX: D4 on Minima, RXD13, TX:D10 on WiFi, **Note**: external transceiver required)
- GND
- AREF (analog reference voltage, connected to +5 V through a 5.1-kΩ resistor)
- I^2C SDA (pullups not mounted)
- I^2C SCL (pullups not mounted)

ICSP connector:

- See Table 1.2

SWD/JTAG connector:

- See Table 1.3

Pin	Function	Type	Description
1	CIPO	Internal	Controller In Peripheral Out
2	+5V	Internal	Power Supply of 5 V
3	SCK	Internal	Serial Clock
4	COPI	Internal	Controller Out Peripheral In
5	RESET	Internal	Reset
6	GND	Internal	Ground

Table 1.2: ICSP connector pins (source: Product Reference Manual, SKU: ABX00080).

Pin	Function	Type	Description
1	+5V	Internal	Power Supply of 5 V
2	SWDIO	Internal	Data I/O pin
3	GND	Internal	Ground
4	SWCLK	Internal	Clock Pin
5	GND	Internal	Ground
6	NC	Internal	Not connected
7	RX	Internal	Serial Receiver
8	TX	Internal	Serial Transmitter
9	GND	Internal	Ground
10	NC	Internal	Not connected

Table 1.3: SWD/JTAG connector pins (source: Product Reference Manual, SKU: ABX00080).

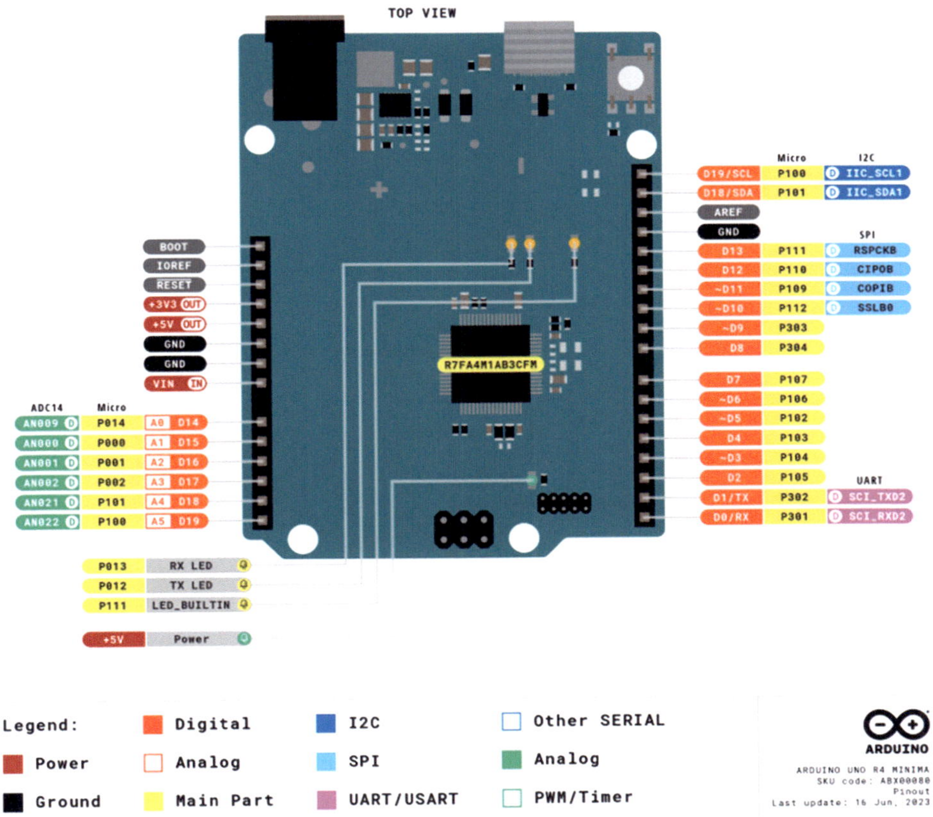

Figure 1.2: Header pins (source: Product Reference Manual, SKU: ABX00080).

The recommended operating conditions are shown in Table 1.4. Schottky diodes are used for overvoltage and reverse polarity protection. Power can either be supplied via the VIN pin, the barrel jack (DC jack), or via USB-C connector. If power is supplied via VIN, a buck converter steps the voltage down to 5 V. Power via USB supplies about 4.7 V (due to Schottky barrier forward bias) to the RA4M1 microcontroller. The RA4M1 processor can operate from +1.6 to +5.5 V and is connected to +5 V on the Arduino board. **The digital GPIOs on the RA4M1 microcontroller can handle currents of IOH = 4 mA and IOL = 8 mA (assuming middle pin drive) current. Remember that IOL is the current into the pin (sinking) when the pin is at logic 0, and IOH is the current from the pin (sourcing) when the pin is at logic 1. Care must be taken not to exceed the recommended current drives of the GPIO ports.** Figure 1.3 shows a simplified power supply connection of the Arduino Uno R4.

Symbol	Description	Min	Typ	Max	Unit
V_{IN}	Input voltage from VIN pad / DC Jack	6	7.0	24	V
V_{USB}	Input voltage from USB connector	4.8	5.0	5.5	V
T_{OP}	Operating Temperature	-40	25	85	°C

Table 1.4: Recommended operating conditions (source: Product Reference Manual, SKU: ABX00080).

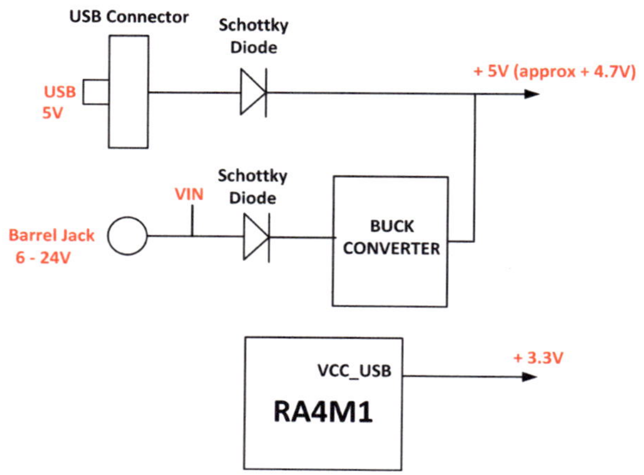

Figure 1.3: Simplified power supply connections.

Figure 1.4 shows the pin layout of the Arduino Uno R4 Minima board. The component layout is shown in Figure 1.5 (taken from Product Reference Manual: SKU: ABX00080) with the component descriptions shown in Table 1.5.

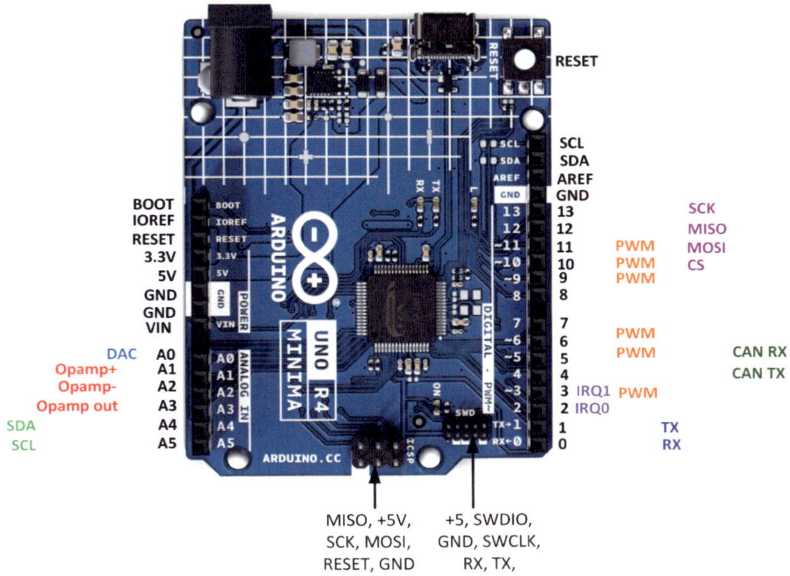

Figure 1.4: Arduino Uno R4 Minima pin layout.

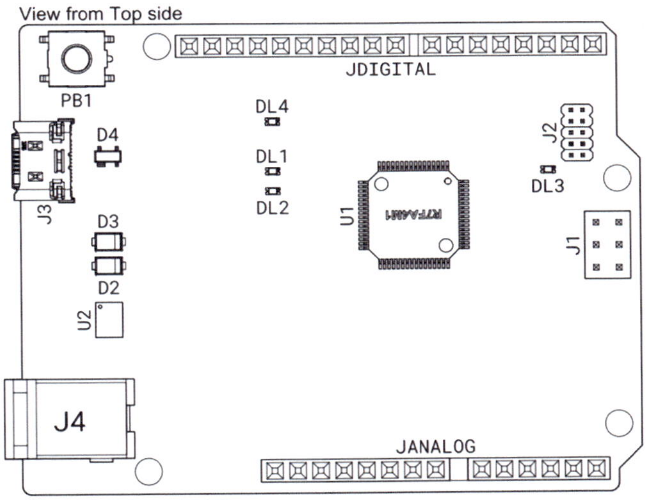

Figure 1.5: Arduino Uno R4 Minima component layout.

Ref.	Description	Ref.	Description
U1	R7FA4M1AB3CFM#AA0 Microcontroller IC	J4	DC Jack
U2	ISL854102FRZ-T Buck Converter	DL1	LED TX (serial transmit)
PB1	RESET Button	DL2	LED RX (serial receive)
JANALOG	Analog input/output headers	DL3	LED Power
JDIGITAL	Digital input/output headers	DL4	LED SCK (serial clock)
J1	ICSP header (SPI)	D2	PMEG6020AELRX Schottky Diode
J2	SWD/JTAG Connector	D3	PMEG6020AELRX Schottky Diode
J3	CX90B-16P USB-C® connector	D4	PRTR5V0U2X,215 ESD Protection

Table 1.5: Arduino Uno R4 Minima component layout.

1.4 The Arduino Uno R4 Projects Kit

The Arduino Uno Experimenting (SKU 20339) Kit available from Elektor (www.elektor.com) includes a large number of sensors, actuators, buttons, LEDs, plus a breadboard, stepper motor, jumper wires etc. In detail, the kit includes the following components:

- 1× RFID reader module
- 1× DS1302 clock module
- 1× 5 V stepper motor
- 1× Stepper motor 2003 drive board
- 5x Green Led
- 5× Yellow LED
- 5× Red LED
- 2× Rocker switch
- 1× Flame sensor
- 1× LM35 sensor module
- 1× Infrared receiver
- 3× Light-dependent resistor
- 1× Remote controller
- 1× Breadboard
- 4× Pushbutton (with four caps)
- 1× Buzzer
- 1× Piezo sounder
- 1× Adjustable resistor
- 1× 74HC595 shift register
- 1× 7-segment display
- 1× 4-digit 7-segment display
- 1× 8×8 Dot-matrix display
- 1× 1602 / I²C LCD module
- 1× DHT11 Temperature and humidity module
- 1× Relay module
- 1× Sound module
- Set of Dupont cables
- Set of Breadboard cables
- 1× Water sensor
- 1× USB cable

- 1× PS2 Joystick
- 5× 1 kΩ resistor
- 5× 10 kΩ resistor
- 5× 220Ω resistor
- 1× 4×4 keypad module
- 1× SG90 Servo
- 1× RFID card
- 1× RGB module
- 2× jumper cap
- 1× 9 V Battery DC clip-on cable

The kit is supplied in a plastic box with a lid as shown in Figure 1.6 (note: the actual packaging and contents of the kit as received may differ from the photo).

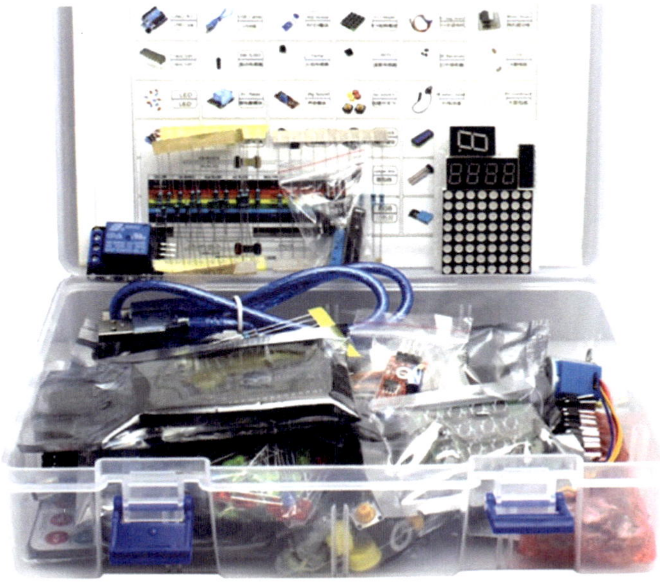

Figure 1.6: The kit supplied in a plastic box.

Figure 1.7 shows the supplied components and the Arduino Uno R4, which is not included and must be purchased separately. A close-up picture of the included sensors, actuators, and displays is shown in Figure 1.8. Figure 1.9 shows a close-up picture of the supplied LEDs, resistors, buttons, buzzers, breadboard and wire jumpers.

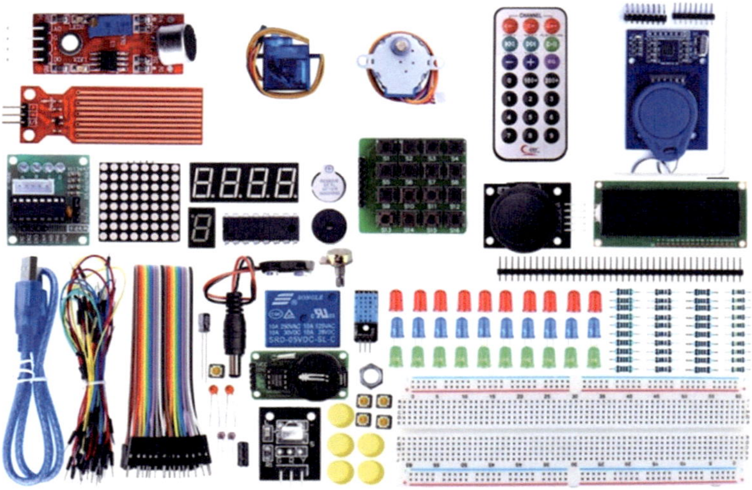

Figure 1.7: Supplied components.

Figure 1.8: Close-up picture of sensors, actuators, and displays.

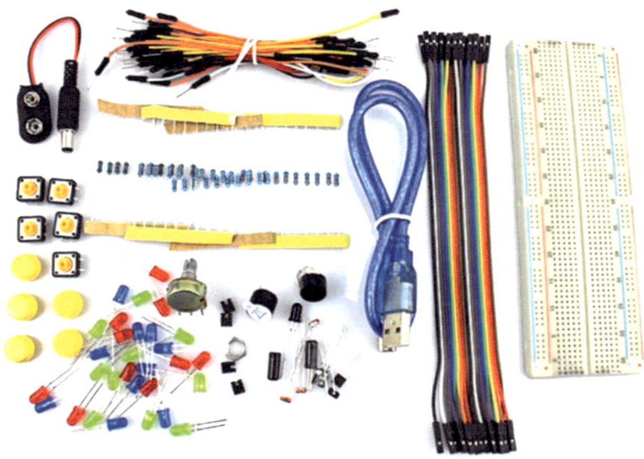

Figure 1.9: Close-up picture of the LEDs, resistors, buttons, wires, etc.

Chapter 2 • Arduino Uno R4 Program Development

2.1 Overview

The Arduino code is written in the C++ language, which is currently one of the most popular languages used to program microcontrollers. Arduino can be programmed using either the **Desktop Arduino IDE** or the **Arduino Web Editor**. The IDE requires the software to be installed on your PC. This is an Integrated Development Environment (IDE) that consists of an editor, compiler, debugger, and tools to upload the compiled code to the processor on the development board. An Arduino program is called a **sketch**, which is compiled into machine code and uploaded to the target processor. Arduino Web Editor enables the programmer to develop, compile, and upload programs using an online browser with the advantage that the online tool is always up-to-date and includes the latest libraries and features. In this book, you will be using the IDE. Interested readers can search for **Arduino Web Editor** in Google and create an account to sign in and use it.

The Arduino IDE has been developed over a decade and there are several versions of it. The latest stable version is **2.1** released in March 2022. Version 1.8.19 has been popular for many years and is still used by many programmers. New version 2.1.0 is the recommended version since it is faster and easier to use than version 1.8.19. The author has used version 2.1.0 in all the projects in this book. Readers may prefer to use the same version in their projects.

In this chapter, you will learn how to install version 2.1.0 of the IDE, which was the latest version at the time of authoring this book. Simple software-only programs are given in this chapter to review the principles of programming using the Arduino IDE. In the next chapters, you will be using the newly released **Arduino Uno R4 Minima/WiFi** development boards together with the supplied components of the bundle in many real-time project applications.

Further information on Arduino IDE, Web Editor, and related tools can be obtained from the links given in Table 2.1.

Ref	Link
Arduino IDE (Desktop)	https://www.arduino.cc/en/Main/Software
Arduino IDE (Cloud)	https://create.arduino.cc/editor
Cloud IDE Getting Started	https://docs.arduino.cc/cloud/web-editor/tutorials/getting-started/getting-started-web-editor
Arduino Project Hub	https://create.arduino.cc/projecthub?by=part&part_id=11332&sort=trending
Library Reference	https://github.com/arduino-libraries/
Online Store	https://store.arduino.cc/

Table 2.1: Links to Arduino IDE, Web Editor, and related tools (Product Reference Manual, SKU: ABX00080).

2.2 Installing the Arduino IDE 2.1.0

The latest version of the Arduino IDE can be installed from the following website:

https://www.arduino.cc/en/software

Select your processor from the **DOWNLOAD OPTIONS** at the right (Figure 2.1). Click **JUST DOWNLOAD** (unless you want to support by contributing). The author installed it on a Windows 10 laptop and at the time of drafting this book the latest version file name was: **Arduino-ide_2.1.0-Windows_64bit.exe**.

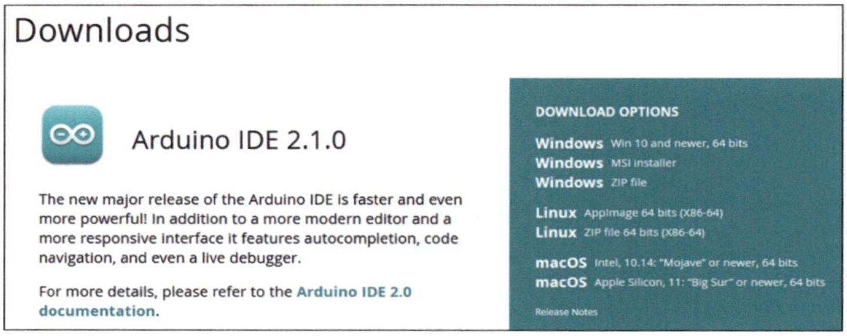

Figure 2.1: Select your processor to install.

You now have to install the Board Package for your Arduino Uno R4 Minima. The steps are:

- Start the Arduino IDE.
- Click to Open the **Boards Manager** at the top left of the screen (Figure 2.2).

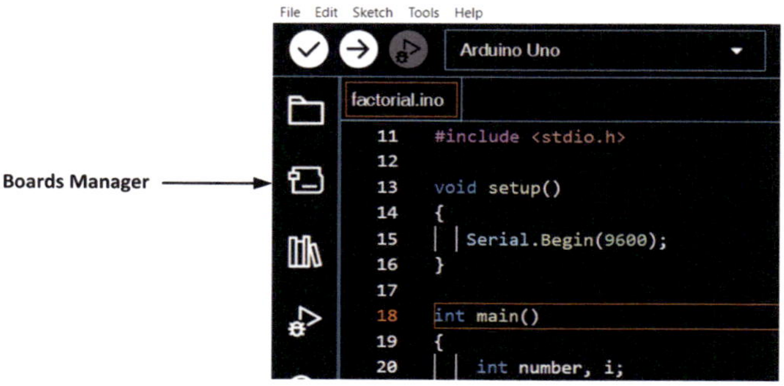

Figure 2.2: Open the Boards manager.

- Search for ARDUINO UNO R4 (Figure 2.3) and click INSTALL to install it. At the time of drafting this book, the version was: 1.0.1.

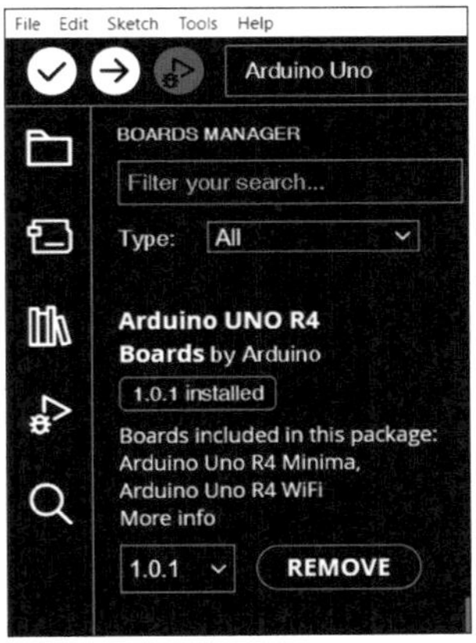

Figure 2.3: Click INSTALL to install the Uno R4 boards.

- Click **Boards Manager** to close the left window.
- You should be able to select the Arduino Uno R4 board from the board selector at the top left of the screen (Figure 2.4). Connect your Arduino Uno R4 Minima (or WiFi) to your PC via a USB-C cable.
- Click **Tools** → **Por**t and select the serial port connected to your Uno R4.

Figure 2.4: Select: Arduino Uno R4 Minima.

You are now ready to develop programs and upload them to your Arduino Uno R4 Minima/ WiFi processor on your development board.

Before looking at some example programs, it is worthwhile to learn some of the *commonly used* menu options offered by the IDE.

File: with this menu option you can open existing programs, save programs, open example programs, and set the IDE working environment (e.g., Preferences...).

Edit: with this menu option you can cut, paste, select, go to a specified line, change indentation and font size, and find text in a file.

Sketch: with this menu option you can compile your program, upload the compiled program to the target processor, include libraries, add files and some other options that you will not be using.

Tools: with this menu option you can manage libraries, configure the serial monitor and serial plotter, select and configure the development board that you will be using, and burn a new bootloader.

Help: this menu option displays various help items on selected topics. Additionally, it displays the version number of the currently used IDE (Figure 2.5). e.g., Version: 2.1.0

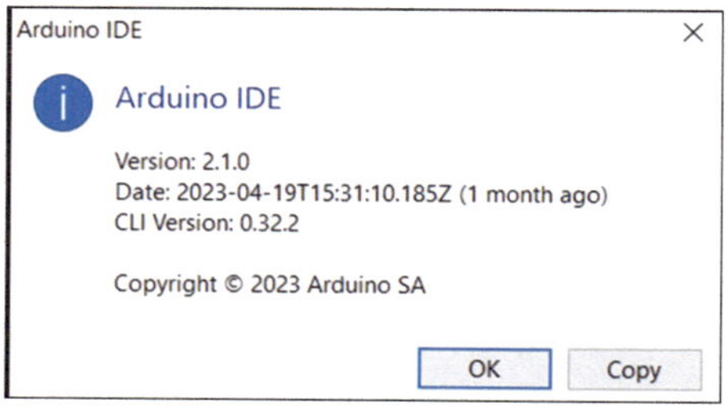

Figure 2.5: Displaying the IDE version number.

2.3 Software-only programs

In this chapter, simple software-only programs are given to review the basic principles of programming in C language using the Arduino IDE. The human interface (e.g. display and keyboard) to these programs is the **Serial Monitor** of the IDE. The aim here has been to review the C language programming concepts by developing simple programs, and then uploading and running them on your development board. Readers who have a good working knowledge of the C language and who are familiar with the Serial Monitor can skip this chapter.

2.3.1 Example 1: Sum of integer numbers

Write a program to read an integer number N from the keyboard and then calculate and display the sum of all the integer numbers from 1 to N.

Solution 1

Figure 2.6 shows the program listing (Program: **sumN**). Comments are used at the beginning of the program to describe the function of the program. Also, the names of the author and the program and the date of development of the program are all listed here. It is strongly recommended by the author to include comments in your programs to make them easy to follow and also easy to modify in the future.

The **setup()** function is executed only once at the beginning of the program. Inside this function, the Serial Monitor is configured to run at 9600 baud (you may choose a different baud rate if you wish).

The main program runs inside the function **loop()**. Here, variables **i**, **N**, and **Sum** are declared as integers and **Sum** is cleared to 0. The program prompts the user to enter **N** which is read using the built-in function **parseInt()**. The program checks if data is available before reading from the keyboard. Then, a **for** loop is formed where the sum of all the integer numbers from 1 to N are calculated and stored in variable **Sum**. The sum is finally displayed as an integer using a **println()** function. Notice that the **println()** function prints a carriage return and line feed after displaying the data. The program is stopped by using a **while()** statement at the end, otherwise the **loop()** function will repeat forever.

The steps to test the program are as follows:

- Connect your Arduino Uno R4 development board to the PC and configure the serial link.
- Type your program as in Figure 2.6 (or load from the Elektor website of the book) and then save it with a suitable name, e.g., **sumN**.
- Click **Sketch → Verify/Compile** to compile the program. The status of the compilation will be displayed in the bottom panel as **Compiling sketch....** If there are any errors, you should go back to your program to correct the errors. If there are no errors, then the bottom panel will display as shown in Figure 2.7 (Click **Output** at the bottom panel to see this message).
- Click **Sketch → Upload** to upload the correctly compiled code to the processor on your development board. You should see the message **Done uploading** at the bottom part of the screen.
- Make sure the Baud rate is set to 9600. Click **Serial Monitor** at the top of the bottom panel. (Figure 2.8). If Serial Monitor is not available, click the **Serial Monitor** icon at the top right corner of the display.

```
//----------------------------------------------------------------
//           SUM OF INTEGER NUMBERS FROM 1 to N
//           ==================================
//
// This program calculates and displays the sum of integer numbers
//
// Author: Dogan Ibrahim
// File  : sumN
// Date  : June, 2023
//----------------------------------------------------------------
void setup()
{
  Serial.begin(9600);
  delay(5000);
}

void loop()
{
  int i, N, Sum = 0;
```

```
  Serial.print("How many numbers are there ? ");
  while(Serial.available() <= 0);
  N = Serial.parseInt();
  Serial.print(N);

  for (i = 1; i <= N; i++)                      // Do for 1 to N
  {
    Sum = Sum + i;                              // Calculate the sum
  }

  Serial.println();                             // Print new line
  Serial.print("Sum of numbers from 1 to ");   // Display "Sum = "
  Serial.print(N);                              // Display N
  Serial.print(" are: ");                       // Display " are: "
  Serial.println(Sum);                          // Display the sum

  while(1);                                     // Stop the program
}
```

Figure 2.6: Program: ***sumN****.*

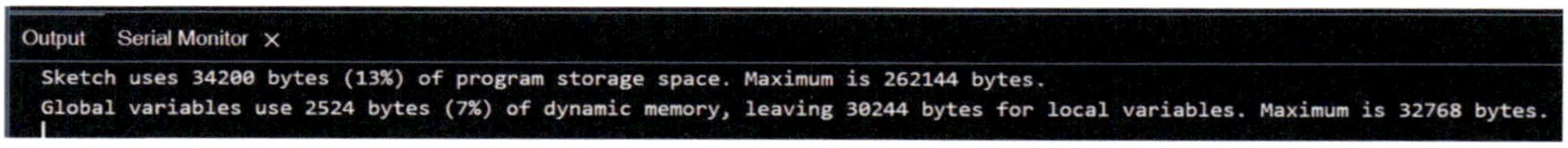

Figure 2.7: Successful compilation.

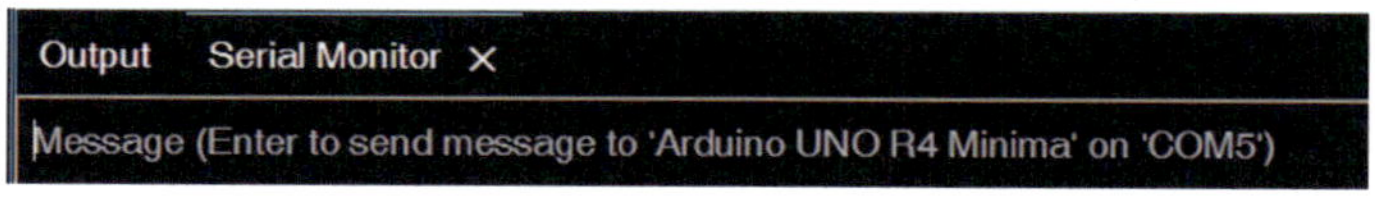

Figure 2.8: Serial Monitor window.

- You should see the text **How many numbers are there ?** displayed at the Serial Monitor window. For example, enter 10 where it says**: Message (Enter to send message to 'Arduino UNO R4' Minima...)** and press the **Enter** key on your PC. You should see 55 displayed which is the sum of integer numbers from 1 to 10 as shown in Figure 2.9.

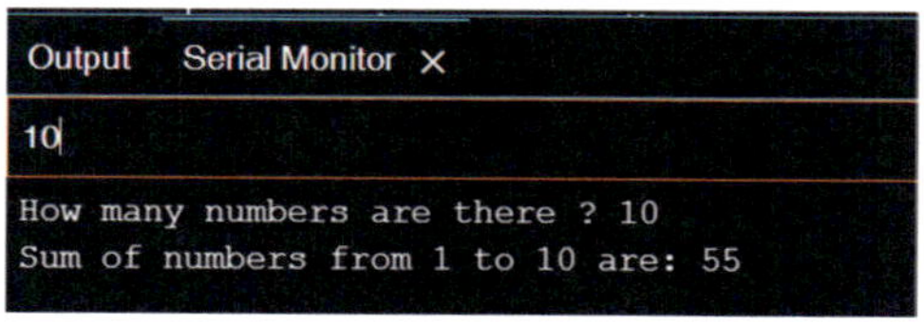

Figure 2.9: Displaying the sum of numbers from 1 to 10.

You should follow the steps given in this program in order to test the other programs given in the following sections of this chapter.

2.3.2 Example 2: Table of squares

Write a program to tabulate the squares of integer numbers from 1 to 10.

Solution 2

Figure 2.10 shows the program listing (Program: **squares**). The Serial Monitor is initialized as in the previous example. A **for** loop is set up in the main program loop, which tabulates the squares of numbers from 1 to 10. The display items are separated with a tab (i.e. "\t").

```
//----------------------------------------------------------------------
//                TABLE OF SQUARES
//                ================
//
// This program displays table of squares of integers from 1 to 10
//
// Author: Dogan Ibrahim
// File  : squares
// Date  : June, 2023
//----------------------------------------------------------------------
void setup()
{
  Serial.begin(9600);
  delay(5000);
}

void loop()
{
  int i, N;

  Serial.println("TABLE OF SQUARES FROM 1 TO 10");
  Serial.println("=============================");
  Serial.println("N\tSQUARE");

  for (i = 1; i <= 10; i++)                  // Do for 1 to N
  {
    N = i * i;                               // Calculate the square
    Serial.print(i);
    Serial.print("\t");
    Serial.println(N);
  }

  while(1);                                  // Stop the program
}
```

Figure 2.10: Program: ***squares****.*

Figure 2.11 shows the output from the program, displayed on the Serial Monitor.

```
TABLE OF SQUARES FROM 1 TO 10
=============================
N        SQUARE
1        1
2        4
3        9
4        16
5        25
6        36
7        49
8        64
9        81
10       100
```

Figure 2.11: Output from the program.

2.3.3 Example 3: Volume of a cylinder

Write a function to return the volume of a cylinder where the radius and height should be passed as arguments to the function. Use the function in a program to calculate and display the area of the cylinder whose radius is 10 cm and height 15 cm.

Solution 3

The volume of a cylinder whose radius and height are *r* and *h* respectively is given by:

$$\text{Area} = \pi r^2 h$$

Figure 2.12 shows the program listing (Program: **cylarea**). Function **area** receives the radius and height of the cylinder as floating point numbers and returns the volume to the main program which then displays the volume on the Serial Monitor.

```
//--------------------------------------------------------------------
//                VOLUME OF A CYLINDER
//                ====================
//
// This program calculates and displays the volume of a cylinder
//
// Author: Dogan Ibrahim
// File  : cylarea
// Date  : June, 2023
//--------------------------------------------------------------------
#define pi 3.14159
float r = 10.0;
float h = 15.0;

void setup()
{
  Serial.begin(9600);
```

```
  delay(5000);
}

//
// Function to calculate the volume
//
float volume(float radius, float height)
{
  float vol;
  vol = pi * radius * radius * height;
  return vol;
}

void loop()
{
  float cylinder;

  Serial.println("Volume of a cylinder with r = 10 cm and h = 15 cm");
  Serial.println("===============================================");

  cylinder = volume(r, h);
  Serial.print("Volume = ");
  Serial.print(cylinder);
  Serial.println(" cm3");

  while(1);                                    // Stop the program
}
```

Figure 2.12: Program: **cylarea**.

Figure 2.13 shows the output from the program where the radius and height are set to 10 cm and 15 cm, respectively at the beginning of the program.

```
Volume of a cylinder with r = 10 cm and h = 15 cm
=================================================
Volume = 4712.38 cm3
```

Figure 2.13: Output from the program.

2.3.4 Example 4: Centigrade to Fahrenheit

Write a program to receive the temperature as Centigrade, convert it to Fahrenheit, and then display it on the Serial Monitor. You should read the temperature from the keyboard.

Solution 4

Given the temperature in degrees C, it can be converted into degrees F using the following formula:

$$F = 1.8 \times C + 32$$

Figure 2.14 shows the program listing (Program: **CtoF**). Function **ToF** receives degrees C as its argument, converts it in degrees F and returns to the main program. The temperature in degrees Centigrade is read from the keyboard.

```
//----------------------------------------------------------------------
//                 CELSIUS TO FAHRENHEIT
//                 =====================
//
// This program converts Celsius to Fahrenheit
//
// Author: Dogan Ibrahim
// File  : CtoF
// Date  : June, 2023
//----------------------------------------------------------------------
void setup()
{
  Serial.begin(9600);
  delay(5000);
}

//
// Function to convert Degrees C to Degrees F
//
float ToF(float C)
{
  return (1.8 * C + 32);
}

void loop()
{
  float F;
  int C;

  Serial.println("Enter temperature as Degreec C: ");
  while(Serial.available() <= 0);
  C = Serial.parseInt();                          // Read Degrees C

  F = ToF(C);                                     // COnvert to F
  Serial.print(C);
  Serial.print(" Degreec C = ");
```

```
  Serial.print(F);
  Serial.println(" Degrees F");

  while(1);                                    // Stop the program
}
```

Figure 2.14: Program: ***CtoF****.*

Figure 2.15 shows the output from the program where 100 degrees centigrade is converted into Fahrenheit and displayed on the Serial Monitor.

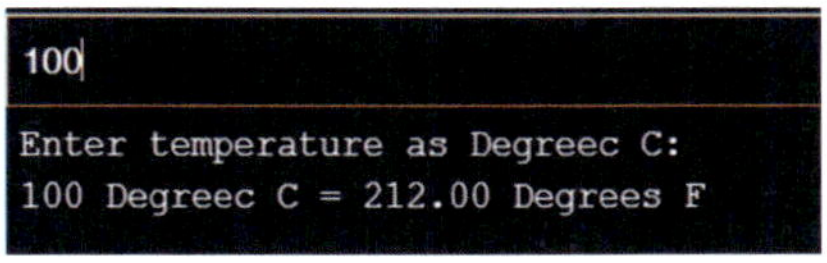

Figure 2.15: Output from the program.

2.3.5 Example 5: Times table

Write a program to read an integer number and then tabulate the times table from 1 to 12 for the given number.

Solution 5

Figure 2.16 shows the program listing (Program: **times**). An integer number is read from the keyboard and stored in variable **N**. Then a **for** loop is set up that runs from 1 to 12. Number **N** is multiplied by 1 to 12 inside this loop and is then displayed on the Serial Monitor.

```
//-----------------------------------------------------------------------
//                          TIMES TABLE
//                          ===========
//
// This program reads an integer number and then tabulates the time table
//
// Author: Dogan Ibrahim
// File  : times
// Date  : June, 2023
//-----------------------------------------------------------------------
void setup()
{
  Serial.begin(9600);
  delay(5000);
}

void loop()
{
  int N, i;
```

```
    Serial.print("Enter the number: ");          // Prompt for a number
    while(Serial.available() <= 0);
    N = Serial.parseInt();                        // Read a number
    Serial.println(N);                            // Display the number

    Serial.print("\nTimes table for number ");   // Heading
    Serial.println(N);

    for(i = 1; i <=12; i++)                       // Do 12 times
    {
      Serial.print(i);                            // Display 1..12
      Serial.print(" X ");                        // Display " X "
      Serial.print(N);                            // Display N
      Serial.print(" = ");                        // Display " = "
      Serial.println(i * N);                      // Dislay the product
    }

    while(1);                                     // Stop the program
}
```

Figure 2.16: Program: ***times****.*

Figure 2.17 shows the output from the program where the times table for number 5 is displayed.

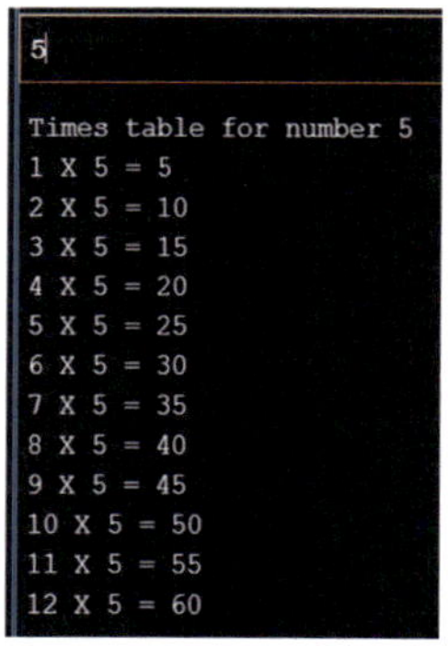

Figure 2.17: Output from the program.

2.3.6 Example 6: Table of trigonometric sine

Write a program to tabulate the trigonometric sine between the angles of 0 to 90 degrees, in steps of 5 degrees.

Solution 6

Figure 2.18 shows the program listing (Program: **sines**). After displaying a heading, a **for** loop is set up which runs from 0 to 90 in steps of 5. Inside this loop, the sine of the angles are calculated. Notice that the angles for trigonometric functions must be entered in radians. Degrees are converted into radians by multiplying with Pi/180.

```
//----------------------------------------------------------------------
//                        TRIGONOMETRIC SINE
//                        ==================
//
// This program displays trigonometric sine from 0 to 90 degrees
//
// Author: Dogan Ibrahim
// File  : sines
// Date  : June, 2023
//----------------------------------------------------------------------
#define pi 3.14159

void setup()
{
  Serial.begin(9600);
  delay(5000);
}

void loop()
{
  int degree;
  float rad;

  Serial.println("TRIGONOMETRIC SINE");             // Heading
  Serial.println("==================");
  Serial.println("Degree\tSine");

  for(degree = 0; degree <=90; degree += 5)       // 0 to 90
  {
    Serial.print(degree);                         // Display degree
    Serial.print("\t");                           // Tab
    rad = degree * pi / 180.0;                    // Convert to radians
    Serial.println(sin(rad));                     // Display sine
  }

  while(1);                                       // Stop the program
}
```

Figure 2.18: Program: **sines.**

Figure 2.19 shows the output from the program.

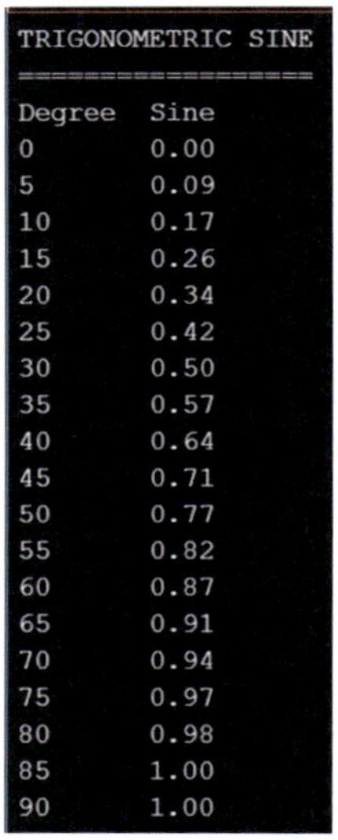
```
TRIGONOMETRIC SINE
==================
Degree  Sine
0       0.00
5       0.09
10      0.17
15      0.26
20      0.34
25      0.42
30      0.50
35      0.57
40      0.64
45      0.71
50      0.77
55      0.82
60      0.87
65      0.91
70      0.94
75      0.97
80      0.98
85      1.00
90      1.00
```

Figure 2.19: Output from the program.

2.3.7 Example 7: Table of trigonometric sine, cosine and tangent

Write a program to tabulate the trigonometric sine between the angles of 0 to 45 degrees, in steps of 5 degrees.

Solution 7

The program is similar to the one given in Figure 2.18, but here cosine and tangent are also included. Figure 2.20 shows the program listing (Program: **trigs**).

```
//----------------------------------------------------------------------
//                   TRIGONOMETRIC SINE,COSINE,TANGENT
//                   =================================
//
// This program displays trigonometric sine,cosine,tangent
//
// Author: Dogan Ibrahim
// File  : trigs
// Date  : June, 2023
//----------------------------------------------------------------------
#define pi 3.14159

void setup()
{
  Serial.begin(9600);
  delay(5000);
}

void loop()
```

```
{
  int degree;
  float rad;

  Serial.println(«TRIGONOMETRIC FUNCTIONS»);      // Heading
  Serial.println(«=======================»);
  Serial.println(«Degree\tSine\tCosine\tTangent»);

  for(degree = 0; degree <=45; degree += 5)       // 0 to 45
  {
    Serial.print(degree);                         // Display degree
    Serial.print(«\t»);                           // Tab
    rad = degree * pi / 180.0;                    // Convert to radians
    Serial.print(sin(rad));                       // Display sine
    Serial.print(«\t»);                           // Tab
    Serial.print(cos(rad));                       // Display cosine
    Serial.print(«\t»);                           // Tab
    Serial.println(tan(rad));                     // Display tangent
  }

  while(1);                                       // Stop the program
}
```

Figure 2.20: Program: ***trigs****.*

Figure 2.21 shows the output from the program.

```
TRIGONOMETRIC FUNCTIONS
=======================
Degree  Sine    Cosine  Tangent
0       0.00    1.00    0.00
5       0.09    1.00    0.09
10      0.17    0.98    0.18
15      0.26    0.97    0.27
20      0.34    0.94    0.36
25      0.42    0.91    0.47
30      0.50    0.87    0.58
35      0.57    0.82    0.70
40      0.64    0.77    0.84
45      0.71    0.71    1.00
```

Figure 2.21: Output from the program.

2.3.8 Example 8: Integer calculator

Write a calculator program. The program should receive two integer numbers from the keyboard and the operation to be performed. The result of the calculation should be displayed on the Serial Monitor. Only the basic four operations (+ – * /) should be used in the program.

Solution 8

Figure 2.22 shows the program listing (Program: **calc**). Two numbers are read from the keyboard and stored in variables **N1** and **N2**. In this program, function **GetIntNumber()** is used to read an integer number from the keyboard. Built-in function **intParse()** could also be used but the problem with **intParse()** function is that it returns 0 the second time it is called because of the carriage return and line feed characters entered after the first call. Here, character array **ibuffer** reads data from the keyboard, converts them into asci by calling built-in function **atoi()** and then returns the integer number to the calling program. Then, the required operation is read by calling function **read()** and is stored in variable **oper**. A **switch** block is used to determine the type of operation to be performed. For example, if **oper** is equal to **'+'** then numbers **N1** and **N2** are added together. The result of the operation is stored in variable **result** which is displayed at the end of the program.

```
//----------------------------------------------------------------------
//                          CALCULATOR
//                          ==========
//
// This is a calculator program that can perform four functions: + - / *
//
// Author: Dogan Ibrahim
// File  : calc
// Date  : June, 2023
//----------------------------------------------------------------------
void setup()
{
  Serial.begin(9600);
  delay(5000);
}

//
// Function to read an integer number
//
int GetIntNumber()
{
  char ibuffer[16];
  int N;
  while(Serial.available() <= 0);
  Serial.readBytes(ibuffer, sizeof(ibuffer));
  N = atoi(ibuffer);
  return N;
}

void loop()
{
  int N1, N2;
  char oper;
```

```
  int result;

  Serial.print("Enter the first number: ");         // Prompt for first number
  N1 = GetIntNumber();                              // Get first number
  Serial.println(N1);                               // Display first number

  Serial.print("Enter the second number: ");        // Prompt for second number
  N2 = GetIntNumber();                              // Get second number
  Serial.println(N2);                               // Display second number

  Serial.print("Enter the operation (+ - * /): "); // Prompt for operation
  while(Serial.available() <= 0);
  oper = Serial.read();                             // Get operation
  Serial.print(oper);                               // Display operation

  switch(oper)
  {
    case '+':                                       // Is it +
      result = N1 + N2;                             // Add
      break;
    case '-':                                       // Is it -
      result = N1 - N2;                             // Subtract
      break;
    case '*':                                       // Is it *
      result = N1 * N2;                             // Multiply
      break;
    case '/':                                       // IS it /
      result = N1 / N2;                             // Divide
      break;
  }

  Serial.print("\nResult = ");                      // Heading
  Serial.println(result);                           // Display result

  while(1);                                         // Stop the program
}
```

Figure 2.22: Program: **calc.**

Figure 2.23 shows an example output from the program where numbers 23 and 3 are multiplied to give the result 69.

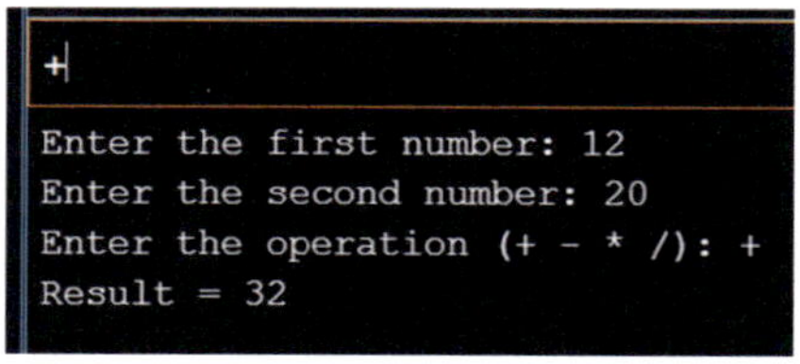

Figure 2.23: Example output from the program.

2.3.9 Example 9: Dice

Write a program to display two random dice numbers between 1 and 6 every time the **Send** button in Serial Monitor is clicked.

Solution 9

Figure 2.24 shows the program listing (Program: **dice**). Inside the **setup()** function, the random number generator seed is initialized with the noise present on analog channel 0. This is necessary so that every time the program is started different set of random numbers are generated (other methods could also be used to initialize the seed). Inside the program loop, two random numbers are generated between 1 and 6 whenever the button of the Serial Monitor is clicked. Clicking this button just sends carriage return and line feed to the program, which are captured by function **readBytes()**. The two random numbers are then generated and displayed on the Serial Monitor. The built-in function **random(min, max)** generates a random integer number between **min** and **max-1**. Notice that the program runs continuously (i.e., it is not stopped at the end as was the case with earlier examples).

```
//-----------------------------------------------------------------------
//                          DICE PROGRAM
//                          ============
//
// This is a dice program that displays two random numbers between 1 and 6
//
// Author: Dogan Ibrahim
// File  : dice
// Date  : June, 2023
//-----------------------------------------------------------------------
void setup()
{
  Serial.begin(9600);
  delay(5000);
  randomSeed(analogRead(0));           // Random noise to get different seed
}

void loop()
{
  int Dice1, Dice2;
  char ibuffer[16];
```

```
  while(Serial.available() > 0)
  {
    Serial.readBytes(ibuffer, sizeof(ibuffer));
    Dice1 = random(1, 7);                  // First number
    Dice2 = random(1, 7);                  // Second number
    Serial.print("\nDice1 = ");            // Display heading
    Serial.println(Dice1);                 // Display first number
    Serial.print("Dice2 = ");              // Display heading
    Serial.println(Dice2);                 // Display second number
  }
}
```

Figure 2.24: Program: ***dice.***

Figure 2.25 shows the output from the program.

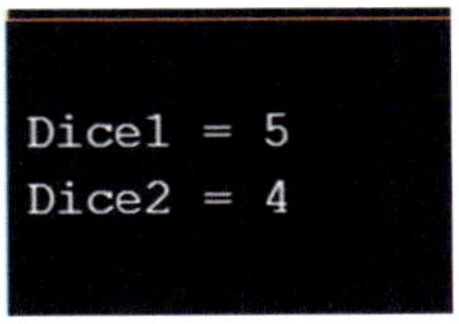

Figure 2.25: Example output from the program.

2.3.10 Example 10: Floating point calculator

This program is similar to the one given in Example 8. Here you want to write a program to operate on integer numbers as well as on floating point numbers.

Solution 10

The program listing is given in Figure 2.26 (Program: **calc2**). The program is very similar to the one given in Figure 2.22, but here the integer declarations are changed to **float**, function **atoi()** is changed to **atof()** and function name is changed to **GetANumber()**.

```
//--------------------------------------------------------------------
//                          CALCULATOR
//                          ==========
//
// This is a calculator program that can perform four functions: + - / *
//
// This is the floating point version of the program
//
// Author: Dogan Ibrahim
// File  : calc2
// Date  : June, 2023
//--------------------------------------------------------------------
void setup()
{
```

```
  Serial.begin(9600);
  delay(5000);
}

//
// Function to read a number
//
float GetANumber()
{
  char ibuffer[16];
  float N;
  while(Serial.available() <= 0);
  Serial.readBytes(ibuffer, sizeof(ibuffer));
  N = atof(ibuffer);
  return N;
}

void loop()
{
  float N1, N2, result;
  char oper;

  Serial.print("Enter the first number: ");          // Prompt for first number
  N1 = GetANumber();                                  // Get first number
  Serial.println(N1);                                 // Display first number

  Serial.print("Enter the second number: ");         // Prompt for second number
  N2 = GetANumber();                                  // Get second number
  Serial.println(N2);                                 // Display second number

  Serial.print("Enter the operation (+ - * /): ");   // Prompt for operation
  while(Serial.available() <= 0);
  oper = Serial.read();                               // Get operation
  Serial.print(oper);                                 // Display operation

  switch(oper)
  {
    case '+':                                         // Is it +
      result = N1 + N2;                               // Add
      break;
    case '-':                                         // Is it -
      result = N1 - N2;                               // Subtract
      break;
    case '*':                                         // Is it *
      result = N1 * N2;                               // Multiply
      break;
```

```
    case '/':                                           // IS it /
      result = N1 / N2;                                 // Divide
      break;
  }

  Serial.print("\nResult = ");                          // Heading
  Serial.println(result);                               // Display result

  while(1);                                             // Stop the program
}
```

Figure 2.26: Program: ***calc2****.*

Figure 2.27 shows an example output from the program where numbers 1.25 and 13.50 are added to give the result 14.75.

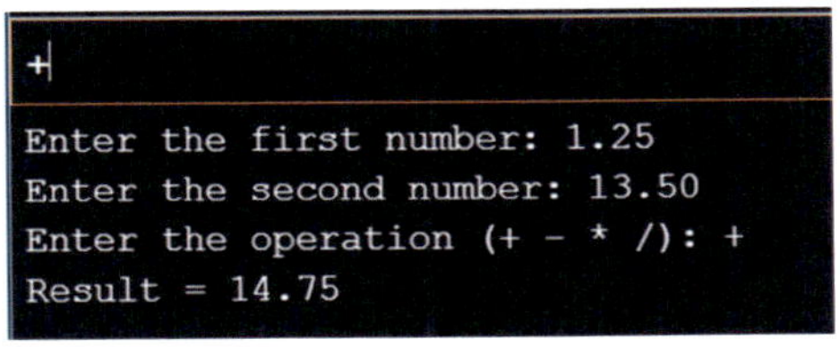

Figure 2.27: Example output from the program.

2.3.11 Example 11: Binary, octal, hexadecimal

Write a program to read a decimal integer number from the keyboard. Convert this number into binary, octal, and hexadecimal and display the results.

Solution 11

Figure 2.28 shows the program listing (Program: **binhexoct**). A decimal integer number is read from the keyboard and is stored in variable **N**. This number is then displayed in binary (**BIN**), octal (**OCT**) and hexadecimal (**HEX**) on the Serial Monitor.

```
//----------------------------------------------------------------------
//               BINARY, HEXADECIMAL,OCTAL
//               =========================
//
// This program reads an integer number and displays it in binary,hex,octal
//
// Author: Dogan Ibrahim
// File  : binhexoct
// Date  : June, 2023
//----------------------------------------------------------------------
void setup()
{
  Serial.begin(9600);
```

```
  delay(5000);
}

void loop()
{
  int N;

  Serial.print("Enter the number: ");          // Prompt for a number
  while(Serial.available() <= 0);
  N = Serial.parseInt();                       // Get the number
  Serial.println(N);

  Serial.print("In binary: ");                 // Display "In binary: "
  Serial.println(N, BIN);                      // Display number in binary
  Serial.print("In octal: ");                  // Display "In octal: "
  Serial.println(N, OCT);                      // Display number in Octal
  Serial.print("In hexadecimal: ");            // Display "In hexadecimal: ")
  Serial.println(N, HEX);                      // Display number in hex

  while(1);                                    // Stop the program
}
```

Figure 2.28: Program: **binhexoct**.

Figure 2.29 shows an example where the decimal number 25 is displayed in binary, octal, and hexadecimal.

```
Enter the number: 25
In binary: 11001
In octal: 31
In hexadecimal: 19
```

Figure 2.29: Example output.

2.3.12 Example 12: String functions

Strings are very important in all programming languages. In C, strings are created either by using NULL terminated character arrays or by using the keyword **String**. There are different functions for either type of string. In this section, you will create various strings and show how to use some of the important string functions.

Solution 12

Figure 2.30 shows the program listing (Program: **strfuncs**). In this program, strings are created as character arrays and also using the **String** keyword. Example output in Figure 2.31 shows the results of the string manipulations.

```
//-----------------------------------------------------------------------
//               STRING FUNCTIONS
//               ================
//
// This program shows how to use various string functions. Both character
// array type and String type strings are used in examples
//
// Author: Dogan Ibrahim
// File  : strfuncs
// Date  : June, 2023
//-----------------------------------------------------------------------
void setup()
{
  Serial.begin(9600);
  delay(5000);
}

void loop()
{
   char mystr[] = "This is a test string";             // create a char array
   char another[80];                                   // Blank char array
   int length;

   // Display the string
   Serial.println(mystr);

   // Display the length of the string (excludes null terminator)
   length = strlen(mystr);
   Serial.print("String length is: ");
   Serial.println(length);

   // Display the length of the string including null terminator
   length = sizeof(mystr);
   Serial.print("Size of the string: ");
   Serial.println(length);

   // Append a string
   strcat(another, " Another string");
   Serial.println(another);

   // Copy a string
   strcpy(another, mystr);
   Serial.println(another);

   // Use the String creation keyword
   String test = "yet another string declaration";
```

```
    // Upper case
    test.toUpperCase();
    Serial.println(test);

    // Display length
    Serial.println(test.length());

    // Add strings
    test = test + " adding";
    Serial.println(test);

    // Position of first character a
    int aposition = test.indexOf('A');
    Serial.println(aposition);

    // Change first character A to X
    test.setCharAt(aposition, 'X');
    Serial.println(test);

    // Replace S with Q
    test.replace('A', 'Q');
    Serial.println(test);

    while(1);                                  // Stop the program
}
```

Figure 2.30: Program: ***strfuncs.***

```
This is a test string
String length is: 21
Size of the string: 22
���� Another string
This is a test string
YET ANOTHER STRING DECLARATION
30
YET ANOTHER STRING DECLARATION adding
4
YET XNOTHER STRING DECLARATION adding
YET XNOTHER STRING DECLQRQTION adding
```

Figure 2.31: Example output.

2.3.13 Example 13: Initializing an array

Write a program to initialize a one-dimensional array with the following values, and then display the values on the Serial Monitor:

2, 4, 5, 8, 10, 25, 100, 280, 34, 22

Solution 13

The program listing is shown in Figure 2.32 (Program: **arrayinit**). Array **MyArray** is initialized inside at the beginning of the program. The elements of the array are then displayed inside the **loop()** function.

```
//----------------------------------------------------------------------
//               INITIALIZE AN ARRAY
//               ===================
//
// This program initializes a one dimensioal array and displays its contents
//
// Author: Dogan Ibrahim
// File  : arrayinit
// Date  : June, 2023
//----------------------------------------------------------------------
int MyArray[10] = {2, 4, 5, 8, 10, 25, 100, 280, 34, 22};

void setup()
{
  Serial.begin(9600);
  delay(5000);

}

void loop()
{
  for (int j = 0; j < 10; j++)
  {
    Serial.print(j);
    Serial.print("\t");
    Serial.println(MyArray[j]);
  }

   while(1);                                  // Stop the program
}
```

Figure 2.32: Program: ***arrayinit.***

Figure 2.33 shows the output from the program.

```
0       2
1       4
2       5
3       8
4       10
5       25
6       100
7       280
8       34
9       22
```

Figure 2.33: Output from the program.

2.3.14 Example 14: Character functions

In this example, you will look at the various character functions.

Solution 14

The program listing is shown in Figure 2.34 (Program: **chfuncs**). These character functions test the type of the given character.

```
//----------------------------------------------------------------------
//               CHARACTER FUNCTIONS
//               ===================
//
// This program shows how to use the character functions in C programs
//
// Author: Dogan Ibrahim
// File  : chfuncs
// Date  : June, 2023
//----------------------------------------------------------------------
void setup()
{
  Serial.begin(9600);
  delay(5000);
}

void loop()
{
  if(isdigit('5'))
    Serial.println("5 is digit");
  else
    Serial.println("5 is not a digit");

  if(isalpha('1'))
    Serial.print("1 is alpha");
```

```
  else
    Serial.println("1 is not alpha");

  if(isalnum('a'))
    Serial.println("a is alphanumeric");
  else
    Serial.println("A is not alphanumeric");

  if(isxdigit('F'))
    Serial.println("F is hexadecimal");
  else
    Serial.println("F is not hexadecimal");

  if(islower('Z'))
    Serial.println("Z is lower case");
  else
    Serial.println("Z is not lower case");

  if(isupper('r'))
    Serial.println("r is upper case");
  else
    Serial.println("r is not upper case");

  if(isspace(' '))
    Serial.println("space");
  else
    Serial.println("Not a space");

   while(1);                                // Stop the program
}
```

Figure 2.34: Program: ***chfuncs****.*

Figure 2.35 gives an example output from the program with the most commonly used character functions.

```
5 is digit
1 is not alpha
a is alphanumeric
F is hexadecimal
Z is not lower case
r is not upper case
space
```

Figure 2.35: Example output.

2.3.15 Example 15: Solution of a quadratic equation

Write a program to calculate the roots of a quadratic equation of the following form and display the roots. Enter *a*, *b*, and *c* from the keyboard.

$ax2 + bx + c = 0$

Solution 15

The roots of a quadratic equation are calculated using the following formula:

$$x1,2 = \frac{-b \pm \sqrt{b^2 - 4a}}{2a}$$

Figure 2.36 shows the program listing (Program: **quadratic**).

```
//--------------------------------------------------------------------
//                         QUADRATIC EQUATION
//                         ==================
//
// This program calculates and displays the roots of a quadratic equation.
// Both real and complex root are calculated and displayed
//
// Author: Dogan Ibrahim
// File  : quadratic
// Date  : June, 2023
//--------------------------------------------------------------------
float a, b, c, root1, root2, rootreal, rootimag;
int real;

void setup()
{
  Serial.begin(9600);
  delay(5000);
}


//
// Function to read a number
//
int GetANumber()
{
  char ibuffer[16];
  float N;
  while(Serial.available() <= 0);
  Serial.readBytes(ibuffer, sizeof(ibuffer));
  N = atof(ibuffer);
  return N;
```

```
}

//
// This function calculates and returns the roots
//
void quadratic()
{
  float det = b * b - 4.0 * a * c;
  if(det >= 0.0)                          // If positive determinant
  {
    real = 1;
    det = sqrt(det);                      // Claculate square root
    root1 = (-b + det) / (2.0 * a);       // Root 1
    root2 = (-b - det) / (2.0 * a);       // Root 2
  }
  else                                    // Negative determinant
  {
    real = 0;
    det = -det;                           // Negate the determinant
    det = sqrt(det);                      // Calculate square toot
    rootreal = -b / (2.0 * a);            // Real part of root
    rootimag = det / (2.0 * a);           // Imaginary part of root
  }
}

void loop()
{
  Serial.print("Enter a: ");              // Enter a
  a = GetANumber();                       // Read a
  Serial.println(a);                      // Display a

  Serial.print("Enter b: ");              // Enter b
  b = GetANumber();                       // Read b
  Serial.println(b);                      // Display b

  Serial.print("Enter c: ");              // Enter c
  c = GetANumber();                       // Read c
  Serial.println(c);                      // Display c

  quadratic();                            // Calculate roots
  if(real == 1)                           // If real roots
  {
    Serial.print("Root1 = ");
    Serial.println(root1, 4);             // Real root 1

    Serial.print("Root2 = ");
```

```
      Serial.println(root2, 4);                 // Real root 2
    }
    else                                        // If complex roots
    {
      Serial.print("Root1 = ");
      Serial.print(rootreal, 4);                // Real root 1
      Serial.print("+j");
      Serial.println(rootimag, 4);              // imaginary root 1

      Serial.print("Root2 = ");
      Serial.print(rootreal, 4);                // Real root 2
      Serial.print("-j");
      Serial.println(rootimag, 4);              // Imaginary root 2
    }

    while(1);                                   // Stop the program
}
```

Figure 2.36: Program: ***quadratic.***

An example output is shown in Figure 2.37 for the solution of the quadratic equation:

$-2x2 + 2x + 1 = 0$

```
Enter a: -2.00
Enter b: 2.00
Enter c: 1.00
Root1 = -0.3660
Root2 = 1.3660
```

Figure 2.37: Solution of: −2x2 + 2x + 1 = 0.

Figure 2.38 shows an example with complex roots for the equation:

$2x2 + 2x + 1 = 0$

```
Enter a: 2.00
Enter b: 2.00
Enter c: 1.00
Root1 = -0.5000+j0.5000
Root2 = -0.5000-j0.5000
```

Figure 2.38: Solution of: 2x2 + 2x + 1 = 0.

2.3.16 Example 16: Lucky day of the week

Write a program to generate a random number between 1 and 7 and then use this number as the lucky day number of the week. Display the day number. Make sure that a new random number set is used each time the program is started.

Solution 16

Figure 2.39 shows the program listing (Program: **lucky**). Inside the **setup()** function, a seed number is generated for the random number generator by adding all the floating values of all the analog channels. This should be more random than using only one analog channel value.

Inside the program loop, a random number is generated between 0 and 10000 and this is modified to be between 1 and 7 so that the generated number is more random than using just the function **random (1, 8)**. Then a **switch** statement is used and the lucky day is displayed depending on the generated random number.

```
//----------------------------------------------------------------------
//                     LUCKY DAY OF THE WEEK
//                     =====================
//
// This program displays your lucky day of the week!
//
// Author: Dogan Ibrahim
// File  : lucky
// Date  : June, 2023
//----------------------------------------------------------------------
void setup()
{
  Serial.begin(9600);
  delay(5000);
  int val = 0;
  for(int i = 0; i < 6; i++)
    val = val + analogRead(i);                // Sum all analogue inputs
  randomSeed(val);                            // Seed for random number
}

void loop()
{
  int day;

  day = random(10000) % 7 + 1;                // Get a number between 1 and 7

//
// Find the lucky day and display on Serial Monitor
//
```

```
  switch(day)
  {
    case 1:
      Serial.println("Your lucky day is MONDAY");
      break;
    case 2:
      Serial.println("Your lucky day is TUESDAY");
      break;
    case 3:
      Serial.println("Your lucky day is WEDNESDAY");
      break;
    case 4:
      Serial.println("Your lucky day is THURSDAY");
      break;
    case 5:
      Serial.println("Your lucky day is FRIDAY");
      break;
    case 6:
      Serial.println("Your lucky day is SATURDAY");
      break;
    case 7:
      Serial.println("Your lucky day is SUNDAY");
      break;
  }

  while(1);                                // Stop the program
}
```

Figure 2.39: Program: **lucky***.*

Figure 2.40 shows an example output from the program.

```
Your lucky day is MONDAY
```

Figure 2.40: Example output.

2.3.17 Example 17: Factorial of a number

Write a program to read an integer positive number from the keyboard and calculate and display its factorial.

Solution 17

Figure 2.41 shows the program listing (Program: **factorial**). The number is read from the keyboard using the function **scanf()**. The factorial of the number is displayed on the screen.

```
//----------------------------------------------------------------------
//                FACTORIAL OF A NUMBER
//                =====================
//
// This program calculates and displays the factorial of a number
//
// Author: Dogan Ibrahim
// File  : factorial
// Date  : June, 2023
//----------------------------------------------------------------------
#include <stdio.h>

void setup()
{
   Serial.begin(9600);
   delay(5000);
}

void loop()
{
    int number, i;
    unsigned long long fact = 1;

    Serial.print("Enter an integer positive number: ");
    while(Serial.available() <= 0);
    number = Serial.parseInt();
    Serial.println(number);

    //
    // Error message if the entered number is negative
    //
    if (number < 0)
        Serial.print("You must enter a positive number!");
    else
    {
        for (i = 1; i <= number; ++i)
        {
            fact = fact * i;
        }
        Serial.print("Factorial of ");
        Serial.print(number);
        Serial.print(" is ");
        Serial.println(fact);
    }
    while(1);
```

```
}
```

Figure 2.41: Program: **factorial.**

Figure 2.42 shows an example output from the program.

```
Enter an integer positive number: 5
Factorial of 5 is 120
```

Figure 2.42: Example output.

2.3.18 Example 18: Add two square matrices

Write a program to add two given square matrices.

Solution 18

Figure 2.43 shows the program listing (Program: **AddMatrices**). The matrices are defined in the **setup()** function. Function **ADDABTOC()** adds two matrices A and B and stores the result in matrix C which is displayed on the screen.

```
//----------------------------------------------------------------------
//                ADD TWO SQUARE MATRICES
//                =======================
//
// This program adds two square matrices A and B and stores the result in
// MAtrix C which is then displayed on the Serial Monitor
//
// Author: Dogan Ibrahim
// File  : AddMatrices
// Date  : June, 2023
//----------------------------------------------------------------------
int i, j, N = 4;
int C[4][4];

int A[4][4] =
    {
      {1, 10, 11, 13},
       {2, 20, 3, 2},
       {13, 12, 2, 4},
       {20, 2, 1, 0}
    };

int B[4][4] =
    {
      {12, 12, 4, 0},
       {1, 22, 5, 1},
       {0, 3, 12, 4},
       {-2, 2, 0, 1}
```

```
    };

//
// Add two square matrices A and B and store in C
//
void ADDABTOC(int A[][4], int B[][4], int C[][4])
{
    int i, j;
    for (i = 0; i < N; i++)
        for (j = 0; j < N; j++)
            C[i][j] = A[i][j] + B[i][j];
}

//
// Displays a matrix in Serial Monitor
//
void DisplayMatrix(int X[][4])
{
    int i, j;
    for (i = 0; i < N; i++) {
        for (j = 0; j < N; j++)
        {
            Serial.print(X[i][j]);
            Serial.print(" ");
        }
        Serial.println("");
    }
}

void setup()
{
   Serial.begin(9600);
   delay(5000);
}


void loop()
{
    Serial.println("Matrix A is: ");
    DisplayMatrix(A);
    Serial.println("");

    Serial.println("Matrix B is: ");
    DisplayMatrix(B);
    Serial.println("");
```

```
    ADDABTOC(A, B, C);

    Serial.println("Matrix C is: ");
    DisplayMatrix(C);

    while(1);
}
```

Figure 2.43: Program: ***AddMatrices****.*

Figure 2.44 shows an example output from the program.

```
Matrix A is:
1 10 11 13
2 20 3 2
13 12 2 4
20 2 1 0

Matrix B is:
12 12 4 0
1 22 5 1
0 3 12 4
-2 2 0 1

Matrix C is:
13 22 15 13
3 42 8 3
13 15 14 8
18 4 1 1
```

Figure 2.44: Example output.

Chapter 3 • Hardware Projects with LEDs

3.1 Overview

In the last chapter, you have learned how to develop software-only projects using the Arduino IDE with your development board. One of the reasons for using the development board is to make hardware-based projects. The optional Arduino Experimenting Kit from Elektor includes many components. In this chapter, you will be developing various projects using the LEDs supplied with the kit. In later chapters and sections, you will be using some of the other components supplied with the kit.

3.2 Project 1: Blinking LED – using the on-board LED

Description: In this project, you will be blinking the on-board LED every second. The aim of this project is to learn how to use the I/O ports of the development board.

Circuit diagram: The on-board LED is connected to port pin 13 of the Arduino Uno R4 board in current sourcing mode through a current-limiting resistor. This means that the LED is ON when logic 1 is applied to its pin.

Program listing: Figure 3.1 shows the program listing (Program: **LEDonboard**). At the beginning of the program, ON and OFF are defined as HIGH and LOW respectively and LED is assigned to port 13. Inside the **setup()** function, the LED is configured as an output. The remainder of the program runs in function **loop()**. Here, the LED is turned ON and OFF with 1-second delay between each output.

```
//----------------------------------------------------------------------
//                        BLINKING THE ONBOARD LED
//                        ========================
//
// This program blinks the onboard LED at port 13 every second
//
// Author: Dogan Ibrahim
// File  : LEDonboard
// Date  : June, 2023
//----------------------------------------------------------------------
#define ON HIGH                                   // Define ON
#define OFF LOW                                   // Define OFF
int LED = 13;                                     // LED at port 13

void setup()
{
  pinMode(LED, OUTPUT);                           // Configure LED as output
}

void loop()
{
  digitalWrite(LED, ON);                          // LED ON
```

```
    delay(1000);                          // 1 second delay
    digitalWrite(LED, OFF);               // LED OFF
    delay(1000);                          // 1 second delay
}
```

Figure 3.1: Program: **LEDonboard***.*

Compile the program and then upload it to the development board. You should see the on-board LED blinking at one-second intervals.

Function **pinMode()** has two arguments: the port number and the mode. The mode can be: INPUT, OUTPUT, or INPUT_PULLUP.

Function **digitalWrite()** has two arguments: port number and value. The value can be HIGH (logic 1) or LOW (logic 0). Note that the analog inputs can also be used as digital pins, referred to as A0, A1, A2 etc.

Function **digitalRead()** has one argument only which is the port number. This function reads digital data (HIGH or LOW) from the specified port. The analog input pins can be used as digital pins, referred to as A0, A1, A2, etc.

3.3 Project 2: Blinking LED – using an external LED

Description: In this project, you will be connecting an external LED to the development board and blink this LED every second. The aim of this project is to show how an external LED can be connected to the development board.

Block diagram: Figure 3.2 shows the block diagram of the project.

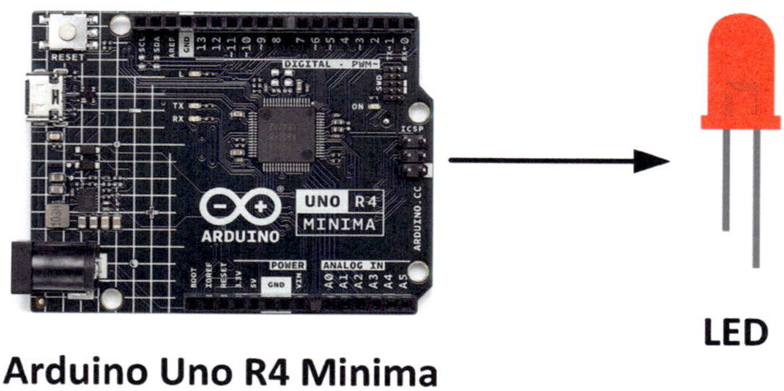

Figure 3.2: Block diagram of the project.

Circuit diagram: LEDs can be connected to the development board as either in current-sourcing mode or current-sinking mode. In current-sourcing mode (Figure 3.3), one pin of the LED is connected to the port through a current-limiting resistor and the other pin is connected to supply ground. In this mode, the LED is ON when logic 1 is applied to it.

In current-sinking mode (Figure 3.4), one pin of the LED is connected to +V power supply and the other pin to the port through a current-limiting resistor. In this mode, the LED is ON when logic 0 is applied to it.

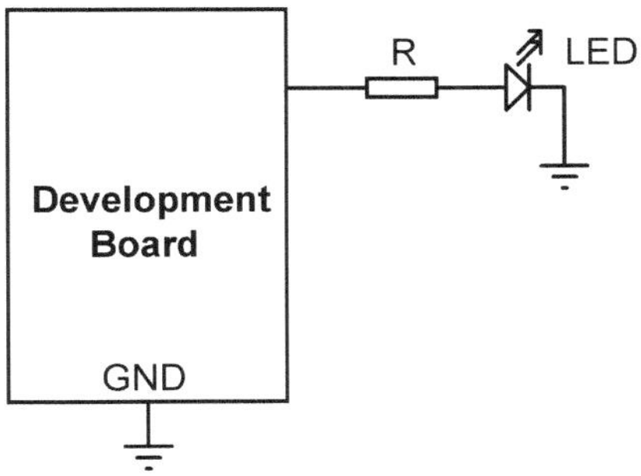

Figure 3.3: Current-sourcing mode.

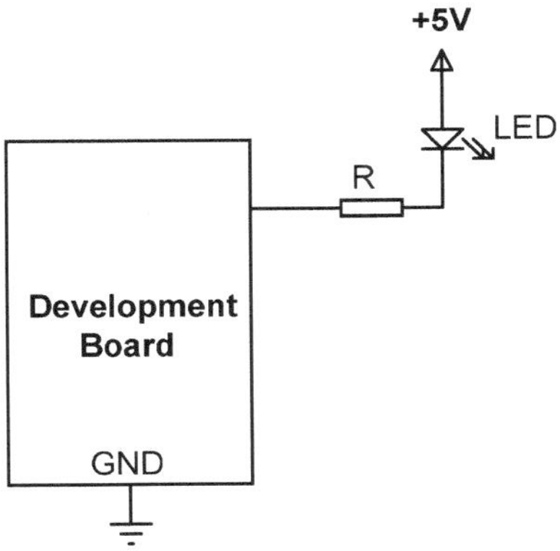

Figure 3.4: Current-sinking mode.

The value of the current-limiting resistor can be calculated as follows: assuming the current through the LED is about 3 mA and the voltage drop across the LED is about 2 V, then

R = (5 V – 2 V) / 3 mA = 1 kΩ. You can use a resistor of around 1 kΩ.

In this project, the LED is connected in current-sourcing mode as shown in Figure 3.5.

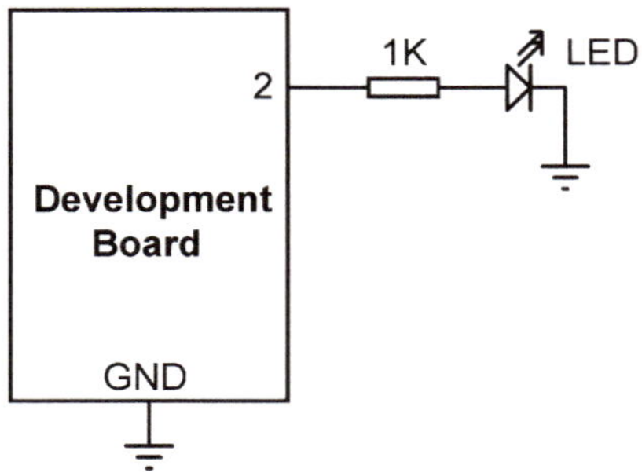

Figure 3.5: Connecting the LED in current-sourcing mode.

Construction: The circuit is constructed on a breadboard and connections are made using jumper wires. Figure 3.6 shows the Fritzing diagram of the circuit. *Note: The pin layout of the Arduino Uno R3 is the same as the board layout of Uno R4.*

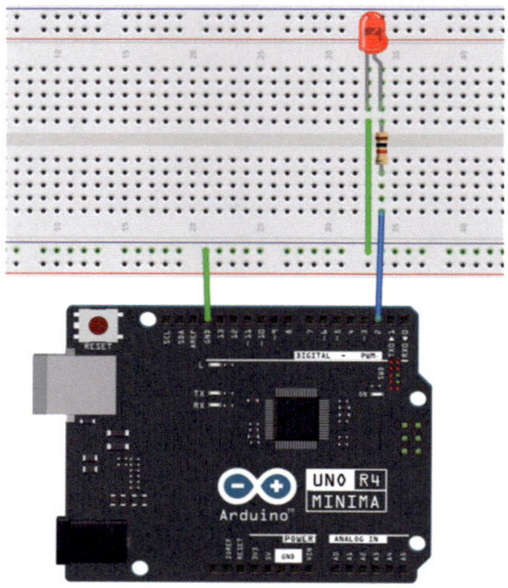

Figure 3.6: Fritzing diagram of the circuit.

Program listing: Figure 3.7 shows the program listing (Program: **LEDext**). The program listing is basically the same as the one given in Figure 3.1, but the LED port is changed from 13 to 2.

```
//------------------------------------------------------------------
//                        BLINKING AN EXTERNAL LED
//                        ========================
//
// This program blinks the LED connected to port 2 of the board
//
// Author: Dogan Ibrahim
// File  : LEDext
```

```
// Date   : June, 2023
//----------------------------------------------------------------------
#define ON HIGH                              // Define ON
#define OFF LOW                              // Define OFF
int LED = 2;                                 // LED at port 2

void setup()
{
  pinMode(LED, OUTPUT);                      // Configure LED as output
}

void loop()
{
  digitalWrite(LED, ON);                     // LED ON
  delay(1000);                               // 1 second delay
  digitalWrite(LED, OFF);                    // LED OFF
  delay(1000);                               // 1 second delay
}
```

Figure 3.7: Program: **LEDext***.*

3.4 Project 3: LED flashing SOS

Description: In this project, an external LED is connected to the development board as in the previous project. The LED blinks in the form of SOS signal (ON ON ON OFF OFF OFF ON ON ON, or in Morse terms: ...---...) with a small delay between each output.

The block diagram, circuit diagram, and the construction of this project are shown in Figure 3.2, Figure 3.5 and Figure 3.6, respectively.

Program listing: Figure 3.8 shows the program listing (Program: **SOS**). The time delay between each **dit** is set to 200 ms, and the time delay between each **dah** is set to 600 ms.

```
//----------------------------------------------------------------------
//                      LED FLASHING SOS
//                      ================
//
// This program blinks the LED connected to port 2 of the board
//
// Author: Dogan Ibrahim
// File  : SOS
// Date  : June, 2023
//----------------------------------------------------------------------
#define ON HIGH                              // Define ON
#define OFF LOW                              // Define OFF
int LED = 2;                                 // LED at port 2
```

```
void setup()
{
  pinMode(LED, OUTPUT);                     // Configure LED as output
}

void loop()
{
  for(int i=0; i < 3; i++)                  // Send S
  {
    digitalWrite(LED, ON);
    delay(200);
    digitalWrite(LED, OFF);
    delay(200);
  }
  delay(500);

    for(int i=0; i < 3; i++)                // Send O
  {
    digitalWrite(LED, ON);
    delay(600);
    digitalWrite(LED, OFF);
    delay(600);
  }
  delay(500);

    for(int i=0; i < 3; i++)                // Send S
  {
    digitalWrite(LED, ON);
    delay(200);
    digitalWrite(LED, OFF);
    delay(200);
  }

  delay(2000);
}
```

Figure 3.8: Program: **SOS***.*

3.5 Project 4: Alternately blinking LEDs

Description: In this project, two LEDs are connected to the development board. The LEDs blink alternately every 500 ms. The aim of this project is to show how multiple LEDs can be connected to the development board.

Block diagram: Figure 3.9 shows the block diagram of the project.

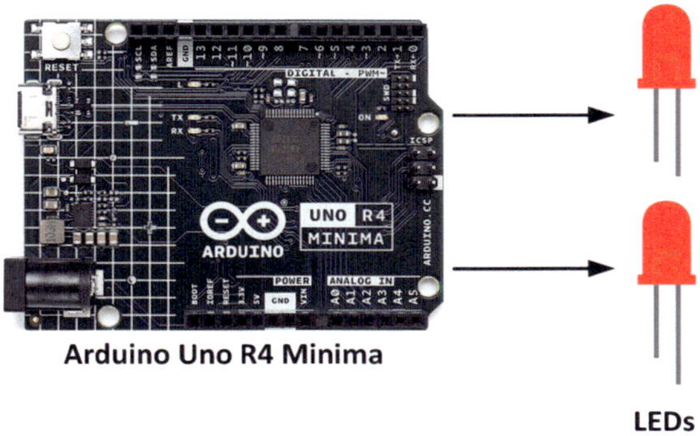

Figure 3.9: Block diagram of the project.

Circuit diagram: In this project, the LEDs are connected in current-sourcing mode as shown in Figure 3.10. The two LEDs are connected to ports 2 and 5 of the development board (you could use any other ports if you wish).

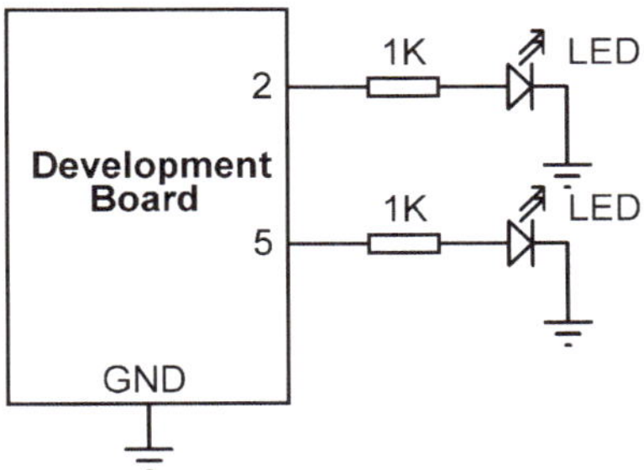

Figure 3.10: Connecting the LEDs in current-sourcing mode.

Construction: The circuit is constructed on a breadboard and connections are made using jumper wires as in the previous project. Figure 3.11 shows the Fritzing diagram of the circuit.

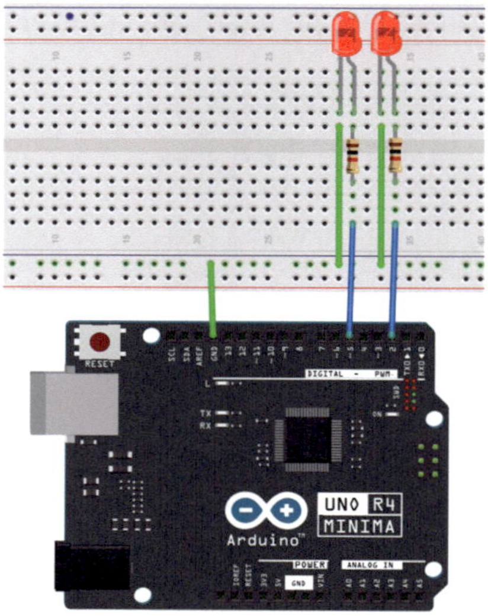

Figure 3.11: Fritzing diagram of the circuit.

Figure 3.12 shows the circuit built on a breadboard.

Program listing: Figure 3.13 shows the program listing (Program: **LEDalternate**). The LEDs are assigned to ports 2 and 5 and are configured as outputs. Inside the main program loop, the LEDs are turned ON and OFF alternately with 500 ms between each output.

```
//------------------------------------------------------------------------
//                        BLINKING ALTERNATE LEDs
//                        =======================
//
// In this program 2 LEDs are connected and they blink alternately
//
```

```
// Author: Dogan Ibrahim
// File  : LEDalternate
// Date  : June, 2023
//----------------------------------------------------------------------
#define ON HIGH                          // Define ON
#define OFF LOW                          // Define OFF
int LED1 = 2;                            // LED1 at port 2
int LED2 = 5;                            // LED2 at port 5

void setup()
{
  pinMode(LED1, OUTPUT);                 // Configure LED1 as output
  pinMode(LED2, OUTPUT);                 // COnfigure LED2 as output
}

void loop()
{
  digitalWrite(LED1, ON);                // LED1 ON
  digitalWrite(LED2, OFF);               // LED2 OFF
  delay(500);                            // 500ms delay
  digitalWrite(LED1, OFF);               // LED1 OFF
  digitalWrite(LED2, ON);                // LED2 ON
  delay(500);                            // 500ms delay
}
```

Figure 3.13: Program: **LEDalternate**.

3.6 Project 5: Chaser-LEDs

Description: In this project, 8 LEDs are connected to the development board. The LEDs "chase" each other as shown in Figure 3.14, with a 500-ms delay between each output.

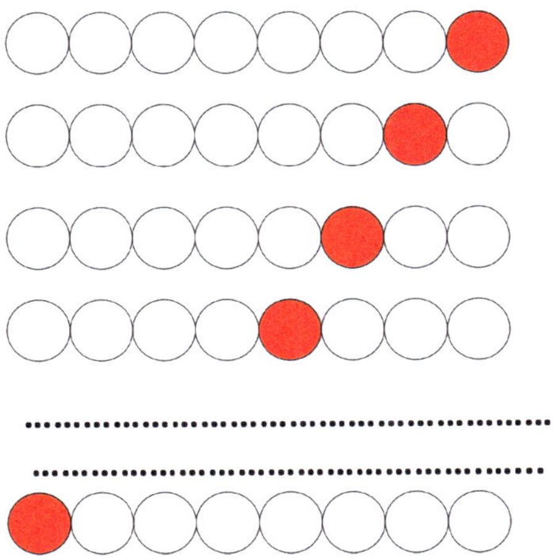

Figure 3.14: Chasing LEDs.

Block diagram: Figure 3.15 shows the block diagram of the project.

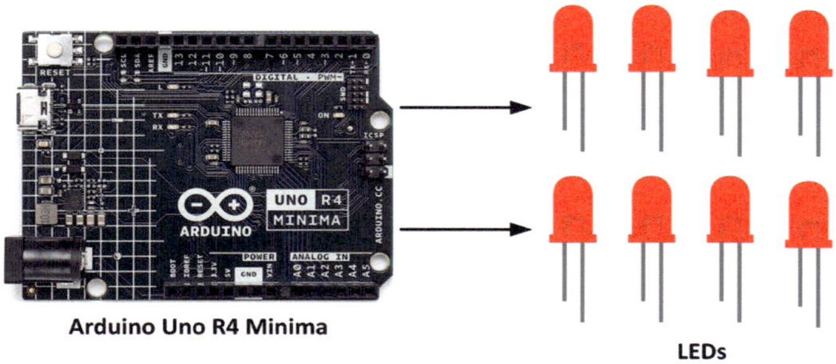

Figure 3.15: Block diagram of the project.

Circuit diagram: Figure 3.16 shows the circuit diagram Ports 2, 3, 4, 5, 6, 7, 8, 9 are connected to LEDs through current-limiting resistors.

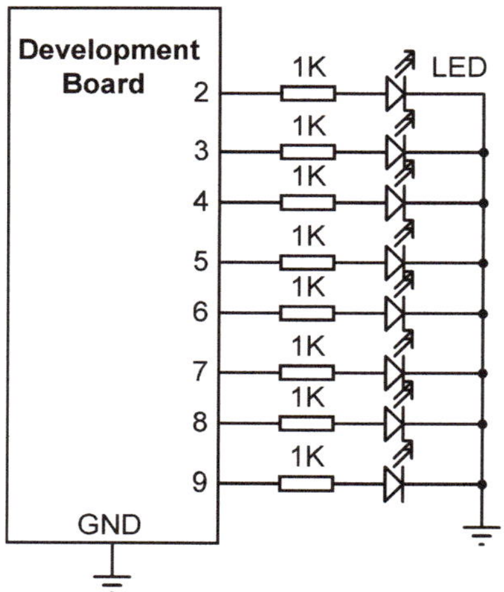

Figure 3.16: Circuit diagram of the project.

Construction: The circuit is constructed on a breadboard and connections are made using jumper wires as in the previous project. Figure 3.17 shows the Fritzing diagram of the circuit.

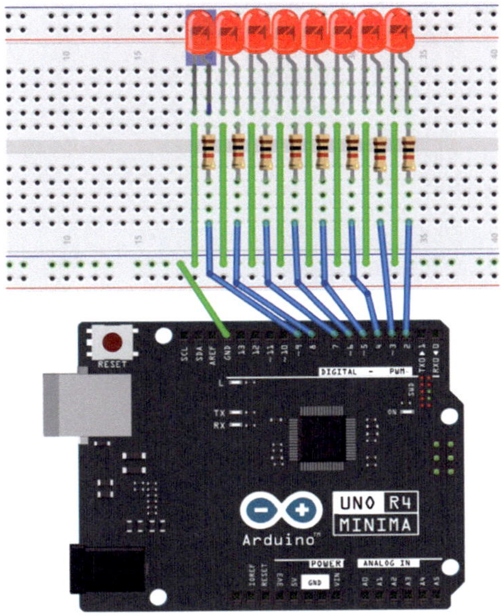

Figure 3.17: Fritzing diagram of the circuit.

Program listing: Figure 3.18 shows the program listing (Program: **LEDchase**). The LEDs are assigned to ports 2 to 9 and are configured as outputs. Inside the main program loop, the LEDs are turned ON and OFF to give the effect of chasing each other as shown in Figure 3.14.

```
//----------------------------------------------------------------------
//                          CHASING 8 LEDs
//                          ==============
//
// In this program 8 LEDs are connected and they chase each other
//
// Author: Dogan Ibrahim
// File  : LEDchase
// Date  : June, 2023
//----------------------------------------------------------------------
#define ON HIGH                                  // Define ON
#define OFF LOW                                  // Define OFF
int LED[] = {2, 3, 4, 5, 6, 7, 8, 9};           // LEDs at ports 2 to 9

void setup()
{
  for(int i = 0; i < 8; i++)
  {
    pinMode(LED[i], OUTPUT);                     // Configure LEDs as outputs
    digitalWrite(LED[i], OFF);                   // LED OFF at beginning
```

```
  }
}

void loop()
{
  for(int i = 0; i < 8; i++)
  {
    digitalWrite(LED[i], ON);                    // LED[i[ ON
    delay(500);                                  // 500 ms delay
    digitalWrite(LED[i], OFF);                   // LED[i] OFF
  }
}
```

Figure 3.18: Program: **LEDchase.**

3.7 Project 6: Chasing LEDs 2

Description: In this project, 8 LEDs are connected to the development board as in the previous project. The LEDs chase each other as shown in Figure 3.19, with a 500-ms delay between each output.

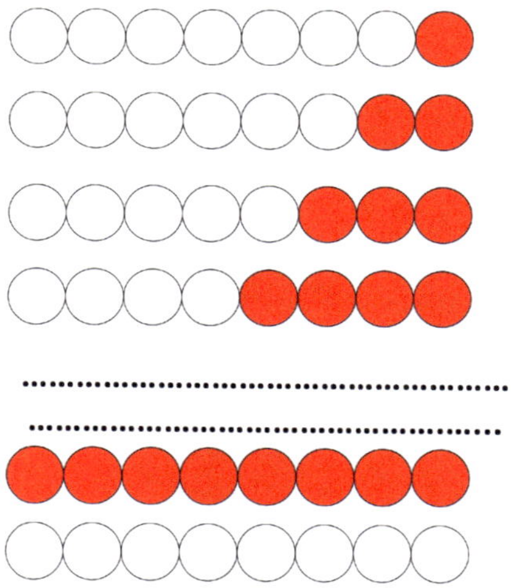

Figure 3.19: Chasing LEDs.

The block diagram, circuit diagram, and Fritzing diagram are shown in Figure 3.15, Figure 3.16 and Figure 3.17, respectively.

Program listing: Figure 3.20 shows the program listing (Program: **LEDchase2**). The LEDs are assigned to ports 2 to 9 and are configured as outputs as in the previous project. Inside the main program loop, the LEDs are turned ON one at a time to give the effect of chasing each other as shown in Figure 3.19. Function **ALLOFF()** turns OFF all the LEDs.

```
//--------------------------------------------------------------------------
//                      CHASING 8 LEDs
//                      ==============
//
// In this program 8 LEDs are connected and they chase each other
//
// Author: Dogan Ibrahim
// File  : LEDchase2
// Date  : June, 2023
//--------------------------------------------------------------------------
#define ON HIGH                             // Define ON
#define OFF LOW                             // Define OFF
int LED[] = {2, 3, 4, 5, 6, 7, 8, 9};       // LEDs at ports 2 to 9

void setup()
{
  for(int i = 0; i < 8; i++)
  {
    pinMode(LED[i], OUTPUT);                // Configure LEDs as outputs
  }
  ALLOFF();                                 // All LEDs OFF
}

//
// Turn OFF all LEDs
//

void ALLOFF()
{
    for(int i = 0; i < 8; i++)
      digitalWrite(LED[i], OFF);              // LEDs OFF at beginning
}

void loop()
{
  for(int i = 0; i < 8; i++)
  {
    digitalWrite(LED[i], ON);               // LED[i[ ON
    delay(500);                             // 500 ms delay
  }
  ALLOFF();                                 // All LEDs OFF
  delay(500);
}
```

Figure 3.20: Program: **LEDchase2***.*

3.8 Project 7: Binary counting LEDs

Description: In this project, 8 LEDs are connected to the development board as in the previous project. As shown in Figure 3.21, the LEDs count up in binary with a 500-ms delay between each count.

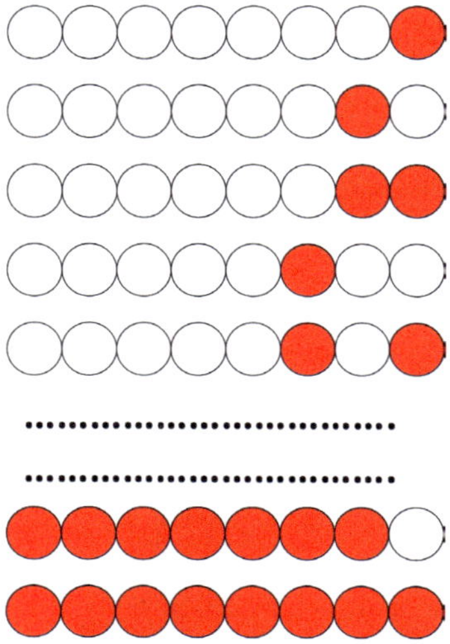

Figure 3.21: Binary counting LEDs.

The block diagram, circuit diagram and the Fritzing diagram are as shown in Figure 3.15, Figure 3.16 and Figure 3.17, respectively.

Program listing: Figure 3.22 shows the program listing (Program: **LEDcount**). The LEDs are assigned to ports 2 to 9 and are configured as outputs in function **setup()**. Function **ALLOFF()** turns OFF all the LEDs. Function **Display()** groups the LEDs as an 8-bit byte and turns ON the LED whose binary value is supplied. Argument **L** is the number of bits in the group (8 here), and **No** is the binary value. For example, function **Display(2, 8)** will turn the second LED from the LSB side ON and all other LEDs will be OFF. Variable **Count** is incremented by one and function **Display()** is called to turn ON the appropriate LEDs so that the LEDs count up in binary as shown in Figure 3.21.

```
//------------------------------------------------------------------
//                        BINARY COUNTING LEDs
//                        ====================
//
// In this program 8 LEDs are connected and they count up in binary
//
// Author: Dogan Ibrahim
// File  : LEDcount
// Date  : June, 2023
//------------------------------------------------------------------
#define ON HIGH                                   // Define ON
```

```
#define OFF LOW                                  // Define OFF
int LED[] = {9, 8, 7, 6, 5, 4, 3, 2};           // LEDs at ports 2 to 9
int Count = 0;

void setup()
{
  for(int i = 0; i < 8; i++)
  {
    pinMode(LED[i], OUTPUT);                     // Configure LEDs as outputs
  }
  ALLOFF();                                      // All LEDs OFF
}


//
// Turn OFF all LEDs
//
void ALLOFF()
{
    for(int i = 0; i < 8; i++)
      digitalWrite(LED[i], OFF);                   // LEDs OFF at beginning
}


//
// Group the port pins together. L is the number of bits (8 here),and No
// is the data to be displayed
//
void Display(int No, int L)
{
  int i, m, j;

  m = L - 1;
  for(i = 0; i < L; i++)
  {
    j = 1;
    for(int k = 0; k < m; k++)j = j * 2;
    if((No & j) != 0)
      digitalWrite(LED[i], ON);
    else
      digitalWrite(LED[i], OFF);
    m--;
  }
}
```

```
void loop()
{
  Count++;                                  // Increment Count
  if(Count > 255) Count = 0;                // If Count>255, set to 0
  Display(Count, 8);                        // Display result
  delay(500);                               // 500 ms delay
}
```

Figure 3.22: Program: ***LEDcount****.*

3.9 Project 8: Random flashing LEDs — Christmas lights

Description: In this project, 8 LEDs are connected to the development board as in the previous project. The lights flash randomly as if they are Christmas lights (more LEDs can easily be added to the project if desired).

The block diagram, circuit diagram, and Fritzing diagram are shown in Figure 3.15, Figure 3.16 and Figure 3.17, respectively.

Program listing: Figure 3.23 shows the program listing (Program: **LEDrandom**). The LEDs are assigned to ports 2 to 9 and are configured as outputs in function **setup()** as in the previous project. A random number is generated between 1 and 255 and this number is used in function **Display()** to turn the LEDs ON/OFF.

```
//----------------------------------------------------------------------
//                        RANDOM FLASHING 8 LEDs
//                        ======================
//
// In this program 8 LEDs are connected and they flash randomly as if
// they are Christmas lights
//
// Author: Dogan Ibrahim
// File  : LEDrandom
// Date  : June, 2023
//----------------------------------------------------------------------
#define ON HIGH                              // Define ON
#define OFF LOW                              // Define OFF
int LED[] = {9, 8, 7, 6, 5, 4, 3, 2};        // LEDs at ports 2 to 9

void setup()
{
  for(int i = 0; i < 8; i++)
  {
    pinMode(LED[i], OUTPUT);                 // Configure LEDs as outputs
  }
}
```

```
//
// Group the port pins together. L is the number of bits (8 here),and No
// is the data to be displayed
//
void Display(int No, int L)
{
  int i, m, j;

  m = L - 1;
  for(i = 0; i < L; i++)
  {
    j = 1;
    for(int k = 0; k < m; k++)j = j * 2;
    if((No & j) != 0)
      digitalWrite(LED[i], ON);
    else
      digitalWrite(LED[i], OFF);
    m--;
  }
}


void loop()
{
  int rnd = random(1, 256);                  // GEenarte random number
  Display(rnd, 8);                           // Display the number
  delay(100);                                // 500 ms delay
}
```

Figure 3.23: Program: **LEDrandom**.

3.10 Project 9: Button controlled LED

Description: In this project, a button and an LED are connected to the development board. The project is very simple: pressing the button turns ON the LED. The aim of this project is to show how a button can be connected and how data can be input.

Block diagram: Figure 3.24 shows the block diagram of the project.

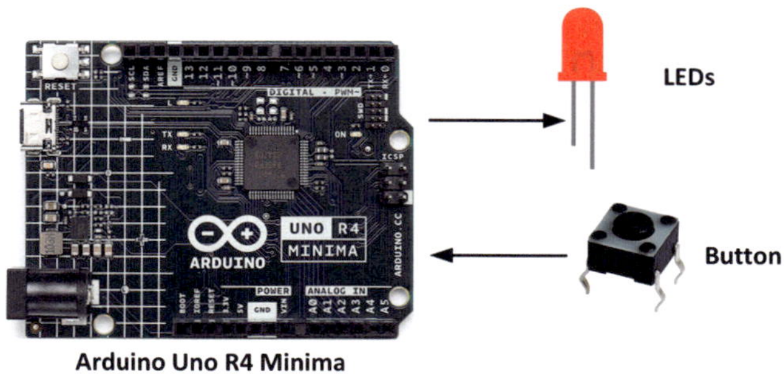

Figure 3.24: Block diagram of the project.

Circuit diagram: A button can be connected in one of two ways. In Figure 3.25, the output state of the button is logic 0 and goes to logic 1 when the button is pressed. In Figure 3.26, the output state of the button is at logic 1 and goes to 0 when the button is pressed.

The circuit diagram of the project is shown in Figure 3.27. The LED is connected to port 2, and the button to port 5. In this project, the output state of the button is at logic 0 and goes to logic 1 when the button is pressed.

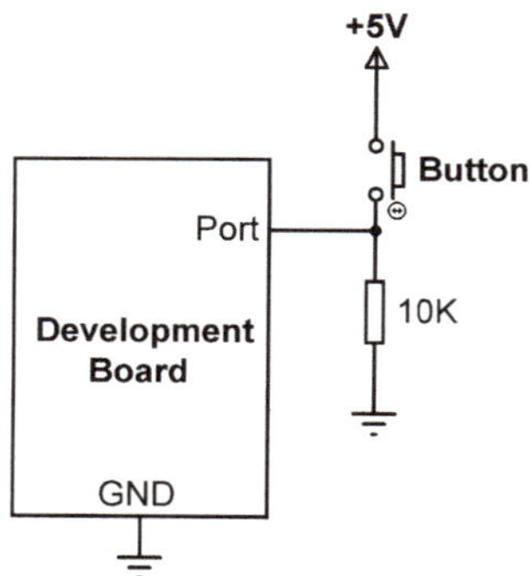

Figure 3.25: Connecting a button.

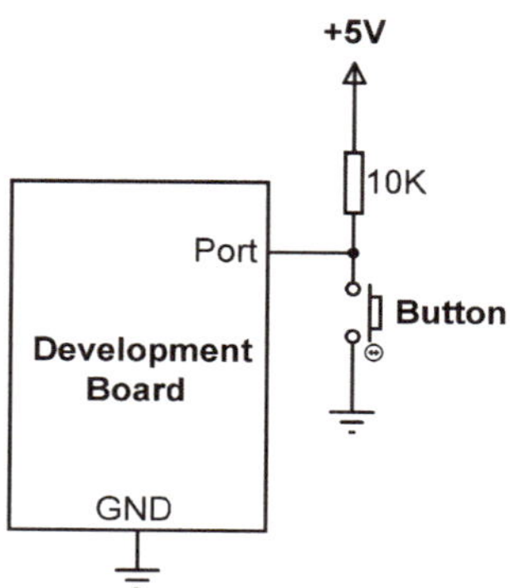

Figure 3.26: Another way of connecting a button.

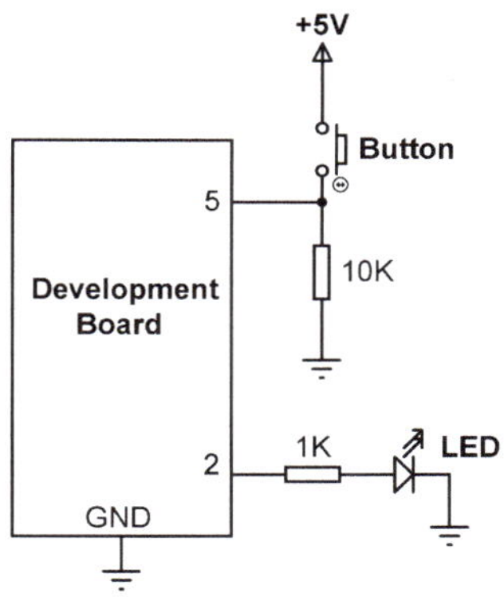

Figure 3.27: Circuit diagram of the project.

Construction: The circuit is constructed on a breadboard and connections made using jumper wires as in the previous projects. Figure 3.28 shows the Fritzing diagram of the circuit.

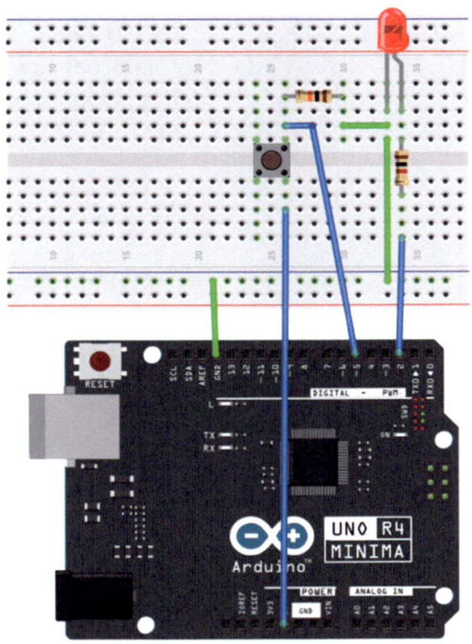

Figure 3.28: Fritzing diagram of the circuit.

Program listing: Figure 3.29 shows the program listing (Program: **Button**). The LED and the button are assigned to ports 2 and 5, respectively. The LED is configured as a digital output, while the button is configured as a digital input. The LED is turned OFF at the beginning of the project. Inside the main program loop, the state of the button is checked. Pressing the button turns ON the LED, and releasing the button turns OFF the LED.

```
//----------------------------------------------------------------------
//                     BUTTON AND LED
//                     ==============
//
// In this program an LED and a button are connected. Pressing the button
// turns ON the LED
//
// Author: Dogan Ibrahim
// File  : Button
// Date  : June, 2023
//-----------------------------------------------------------------------
#define ON HIGH                                 // Define ON
#define OFF LOW                                 // Define OFF
int LED = 2;                                    // LED at port 2
int Button = 5;                                 // Button at port 5

void setup()
{
    pinMode(LED, OUTPUT);                       // Configure LED as output
    digitalWrite(LED, OFF);                     // LED OFF at beginning
    pinMode(Button, INPUT);                     // COnfigure Button as input
}

void loop()
{
  if(digitalRead(Button) == 1)                  // If Button is pressed
    digitalWrite(LED, ON);                      // LED ON
  else                                          // Otherwise
    digitalWrite(LED, OFF);                     // LED OFF
}
```

Figure 3.29: Program: **Button***.*

Modified program

You can pull up the input pin to +V in the software so that you don't have to connect the button to +V. The modified circuit diagram is simpler and is shown in Figure 3.30.

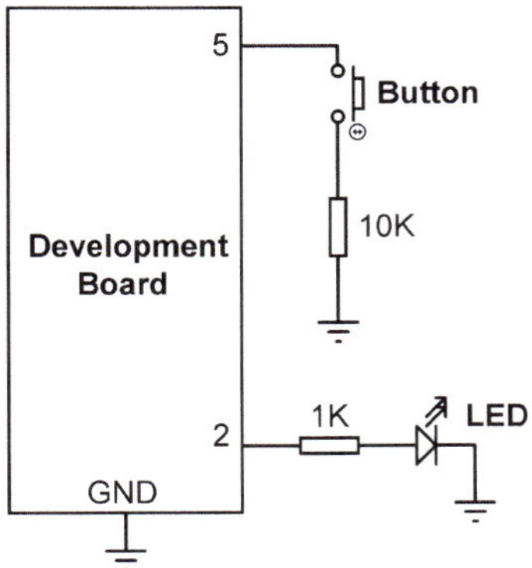

Figure 3.30: Modified circuit diagram.

The modified software (Program: **Button2**) which pulls up the input pin is shown in Figure 3.31. Here, the button pin is normally at logic 1 and goes to logic 0 when the button is pressed.

```
//-----------------------------------------------------------------------
//                       BUTTON AND LED
//                       ==============
//
// In this program an LED and a button are connected. Pressing the button
// turns ON the LED
//
// Modified software that uses software pull-up of input pin
//
// Author: Dogan Ibrahim
// File  : Button2
// Date  : June, 2023
//-----------------------------------------------------------------------
#define ON HIGH                                 // Define ON
#define OFF LOW                                 // Define OFF
int LED = 2;                                    // LED at port 2
int Button = 5;                                 // Button at port 5

void setup()
{
    pinMode(LED, OUTPUT);                       // Configure LED as output
    digitalWrite(LED, OFF);                     // LED OFF at beginning
    pinMode(Button, INPUT_PULLUP);              // Pull-up button pin
}

void loop()
{
  if(digitalRead(Button) == 0)                  // If Button is pressed
```

```
    digitalWrite(LED, ON);                          // LED ON
  else                                              // Otherwise
    digitalWrite(LED, OFF);                         // LED OFF
}
```

Figure 3.31: Program: **Button2**.

3.11 Project 10: Controlling the LED flashing rate — external interrupts

Description: In this project, two buttons named **FAST** and **SLOW** and an LED are used. Pressing FAST will increase the LED flashing rate. Similarly, pressing SLOW will decrease the LED flashing rate. The aim of this project is to show how external interrupts can be used with the Arduino IDE.

Block diagram: Figure 3.32 shows the block diagram of the project.

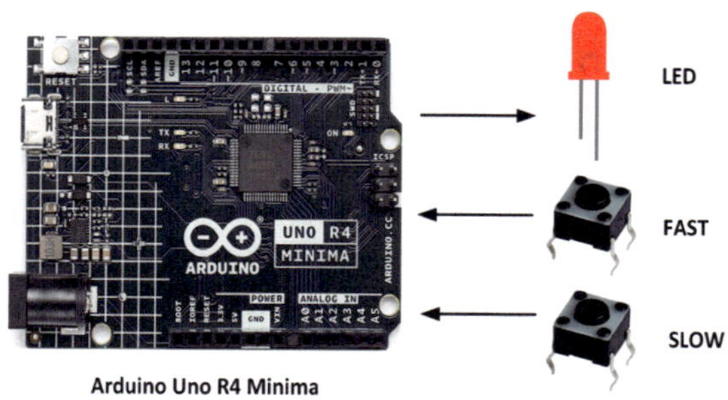

Figure 3.32: Block diagram of the project.

Circuit diagram: As shown in Figure 3.33, the LED is connected to port 5. FAST and SLOW buttons are connected to ports 2 and 3 of the development board. Internal pull-up resistors of the processor are used in the project so that the state of a button is at logic 1 and goes to logic 0 when the button is pressed.

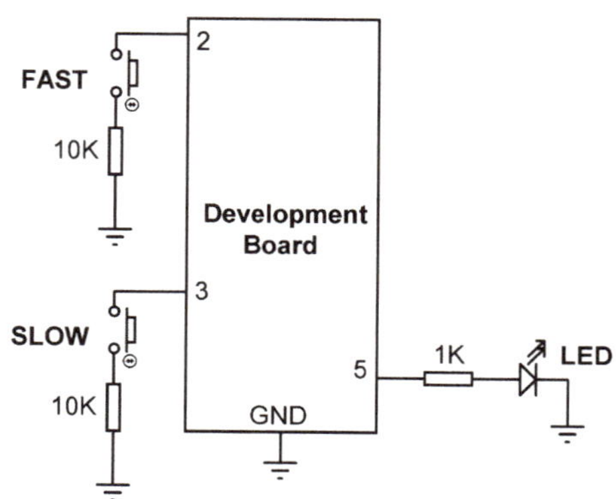

Figure 3.33: Circuit diagram of the project.

Construction: The circuit is constructed on a breadboard and connections made using jumper wires as in the previous projects. Figure 3.4 shows the Fritzing diagram of the circuit.

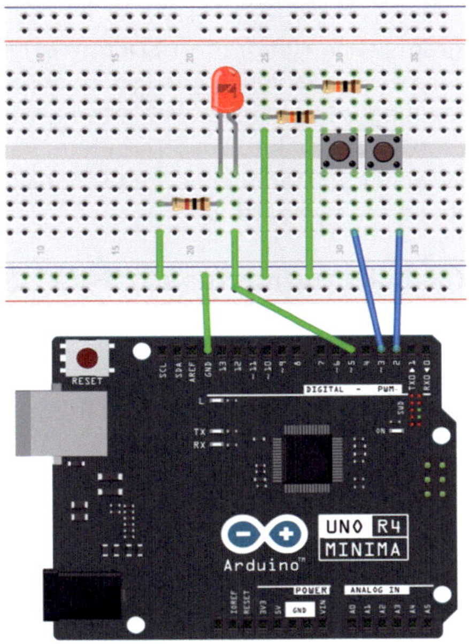

Figure 3.34: Fritzing diagram of the circuit.

Program listing: This program uses external interrupts on the two buttons. When the program is started the LED flashes with one second delay between each output. Pressing the button FAST generates an external interrupts where inside the interrupt service routine this delay is reduced so that the LED flashes faster. Similarly, pressing the button SLOW generates another interrupt where inside the interrupt service routine the delay is increased so that the LED flashes slower.

External interrupts are very important in many microcontroller applications. In this project, because the LED is flashing constantly, you have to use external interrupts to change the flashing rate. This is because you could not test for the button actions inside the flashing code.

Maduino Uno (or Arduino Uno) supports external interrupts on its port pins 2 and 3. The following code must be executed to set up an external interrupt:

attachInterrupt(digitalPinToInterrupt(pin), ISR, mode)

where ISR is the name of the interrupt service routine function name, and mode tells how interrupts will be accepted at the specified pin. Mode can take the following values:

- **LOW**: to trigger the interrupt whenever the pin is LOW.
- **CHANGE**: to trigger the interrupt whenever the pin changes state.
- **RISING**: to trigger the interrupt when the pin goes from LOW to HIGH.
- **FALLING**: to trigger the interrupt whenever the pin goes from HIGH to LOW.

Figure 3.35 shows the program listing (Program: **LEDcontrol**). The LED is at port 5. Buttons are at ports 2 and 3 where external interrupts can be accepted. LED is configured as output, and the buttons are configured as inputs with internal pull-up resistors. Pressing a button changes the button state from High to Low. Therefore, you have to set the interrupt mode to **FALLING**. The following two statements are used inside the **setup()** function to configure external interrupts on the two buttons:

attachInterrupt(digitalPinToInterrupt(2), FAST_CODE, FALLING);
attachInterrupt(digitalPinToInterrupt(3), SLOW_CODE, FALLING);

Where **FAST_CODE** and **SLOW_CODE** are the interrupt service routine function names. The LED flashes every second when the program is started. Pressing the button **FAST** generates an external interrupt and the program jumps to function **FAST_CODE** where the delay is decremented by 50 ms, thus making the LED flash faster. Similarly, pressing the button **SLOW** generates an external interrupt and the program jumps to function **SLOW_CODE** where the delay is increased by 50 ms, thus making the LED flash slower.

```
//----------------------------------------------------------------------
//                      CONTROLLING THE LED FLASHING RATE
//                      =================================
//
// In this program an LED and two buttons named FAST and SLOW are connected.
// Pressing FAST increases flashing rate.Pressing SLOW decreases flashing rate.
//
// The program is based on using external interrupts
//
// Modified software that uses software pull-up of input pin
//
// Author: Dogan Ibrahim
// File  : LEDcontrol
// Date  : June, 2023
//----------------------------------------------------------------------
#define ON HIGH                                     // Define ON
#define OFF LOW                                     // Define OFF
int LED = 5;                                        // LED at port5
int FAST = 2;                                       // FAST button at port 2
int SLOW = 3;                                       // SLOW button at port 3
int dely = 1000;                                    // Default delay (1000 ms)

void setup()
{
```

```
    pinMode(LED, OUTPUT);                       // Configure LED as output
    digitalWrite(LED, OFF);                     // LED OFF at beginning
    pinMode(FAST, INPUT_PULLUP);                // Pull-up FAST button pin
    pinMode(SLOW, INPUT_PULLUP);                // Pull-up SLOW button pin
    attachInterrupt(digitalPinToInterrupt(2), FAST_CODE, FALLING);
    attachInterrupt(digitalPinToInterrupt(3), SLOW_CODE, FALLING);
}

//
// FAST_CODE interrupt service routine
//
void FAST_CODE()
{
  dely = dely - 50;                             // Decrement delay
  if(dely < 0)dely = 0;
}

//
// SLOW_CODE interrupt service routine
//
void SLOW_CODE()
{
  dely = dely + 50;                             // Increment delay
}


void loop()
{
  digitalWrite(LED, ON);                        // LED ON
  delay(dely);                                  // Delay
  digitalWrite(LED, OFF);                       // LED OFF
  delay(dely);                                  // Delay
}
```

Figure 3.35: Program: **LEDcontrol**.

3.12 Project 11: Reaction timer

Description: In this project, an LED and a button are used. The project measures the reaction time of the user and displays it on the Serial Monitor in milliseconds. The LED is turned ON at random times. As soon as the user sees the LED, he/she is expected to press the button. The time delay between seeing the LED and pressing the button is a measure of the reaction time, which is displayed by the program.

The block diagram and circuit diagram of the project are in Figure 3.24 and Figure 3.27, respectively.

Program listing: Figure 3.36 shows the program listing (Program**: reaction**). Inside the **setup()** function, the Serial Monitor is initialized, the LED is configured as output, and the button is configured as. Inside the main program loop, a random number is generated between 1 and 20 seconds, and the program is configured to wait for some random time before turning ON the LED. At this point, the time is read by calling the built-in function **millis()** and is stored in variable **StartTime**. When the button is pressed, the time is read again and stored in **EndTime**. The difference between the **EndTime** and **StartTime** is the reaction time of the user.

Compile and upload the program to the processor. Then, start the Serial Monitor and press the button as soon as the LED is ON.

```
//---------------------------------------------------------------------
//                    REACTION TIME
//                    =============
//
// This program measures the reaction time of the user and displays it
// on the Serial Monitor in milliseconds.
//
// Author: Dogan Ibrahim
// File  : reaction
// Date  : June, 2023
//---------------------------------------------------------------------
#define ON HIGH                                 // Define ON
#define OFF LOW                                 // Define OFF
int LED = 2;                                    // LED at port 2
int Button = 5;                                 // Button at port 5

void setup()
{
    Serial.begin(9600);
    pinMode(LED, OUTPUT);                       // Configure LED as output
    digitalWrite(LED, OFF);                     // LED OFF at beginning
    pinMode(Button, INPUT_PULLUP);              // Pull-up button pin
    delay(2000);
}


void loop()
{
  int rnd = random(1, 21);                      // Random number 1-20 secs
  delay(rnd);                                   // Random delay
  digitalWrite(LED, ON);                        // LED ON
  float StartTime = millis();                   // Start time
  while(digitalRead(Button) == 1);              // Wait for button press
  float EndTime = millis();                     // End time
```

```
    digitalWrite(LED, OFF);                           // LED OFF
    float ElapsedTime = EndTime - StartTime;          // Elapsed time
    Serial.print("Reaction time (ms) = ");            // Heading
    Serial.println(ElapsedTime, 2);                   // Reaction time

    delay(3000);
}
```

Figure 3.36: Program: **reaction.**

Figure 3.37 shows an example output from the program.

```
Reaction time (ms) = 6849.00
Reaction time (ms) = 786.00
Reaction time (ms) = 4222.00
```

Figure 3.37: Example output.

3.13 Project 12: LED color wand

Description: In this project, an RGB LED module is used to generate different colors of light, just like a color wand. The aim of this project is to show how an RGB LED can be used in a program to generate different colors of light.

Block diagram:

The block diagram of the project is shown in Figure 3.38.

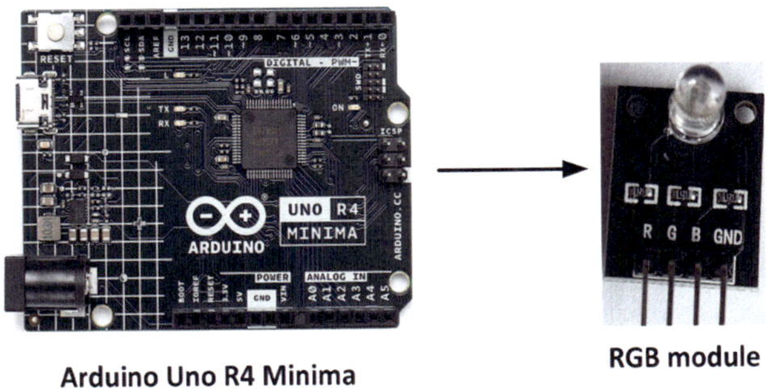

Figure 3.38: Block diagram of the project.

Circuit Diagram: RGB modules can either be common-cathode or common-anode. The one supplied with the kit is common-cathode where the common pin is connected to ground and an LED is turned ON by applying logic 1 to its pin. The module supplied has 4 pins marked R, G, B, and GND. 120-Ω on-board current-limiting resistors are used in series with each pin. With 120 Ω, assuming 2 V drop across each LED, the current drawn by each LED will be about (5 V – 2 V) / 120 Ω or 25 mA, which is well within the I/O specifications (it is recommended to use higher resistor values if higher currents are to be drawn from the I/O ports).

The circuit diagram of the project is shown in Figure 3.39. The R, G, and B LED pins are connected to port pins 2, 3, and 4 of the development board. The common pin of the RGB module is connected to GND.

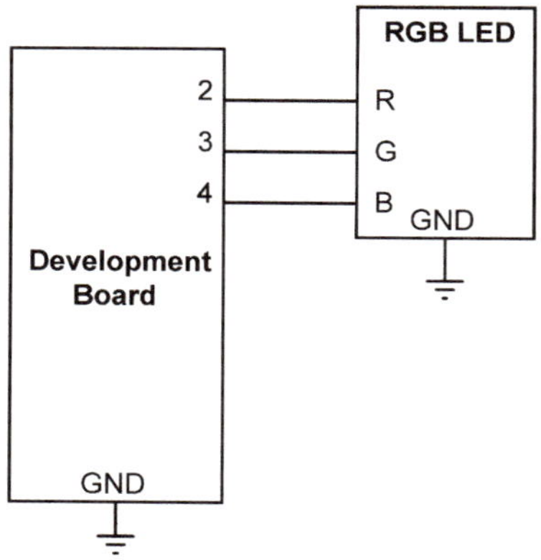

Figure 3.39: Circuit diagram of the project.

Program listing: The program listing of the project is very simple and is shown in Figure 3.40 (Program: **RGB**). At the beginning of the program, the I/O pins where the Red, Green and Blue pins are connected are defined. Then these port pins are configured as outputs. The remainder of the program runs in an endless loop where inside this loop random numbers are generated between 0 and 1 (notice that when using the **random** function, the lower bond is included, but the upper bound is excluded. Because of this the arguments to this function must be 0, 2) for all three colors and the generated numbers are sent to the corresponding ports. Thus, for example, if number 1 is generated for the Red port then the red color is turned ON, and so on.

```
//--------------------------------------------------------------------
//                     LED COLOUR WAND - RGB
//                     =====================
//
// In this project we use an RGB module to generate random colours, just
// like a magic colour wand
//
// Author: Dogan Ibrahim
// File  : RGB
// Date  : June, 2023
//--------------------------------------------------------------------
#define RED 2
#define GREEN 3
#define BLUE 4
int R, G, B;

//
```

```
// Set I/O pin 2, 3 and 4 as outputs
//
void setup()
{
   pinMode(RED, OUTPUT);
   pinMode(GREEN, OUTPUT);
   pinMode(BLUE, OUTPUT);
   randomSeed(10);
}

//
// Use the random number generator to generate ON (1) or OFF
// (0) values for the three colours
//
void loop()
{
   R = random(0, 2);
   G = random(0, 2);
   B = random(0, 2);

   digitalWrite(RED, R);
   digitalWrite(GREEN, G);
   digitalWrite(BLUE, B);
   delay(500);
}
```

Figure 3.40: Program: ***RGB****.*

Suggestion: In the program in Figure 3.40, the delay time is set to 500 ms. Try changing this time (e.g., make it shorter) and see its effects.

3.14 Project 13: RGB fixed colors

Description: In this project, you will generate fixed colors using the RGB module.

The block diagram and circuit diagram of the project are in Figure 3.38 and Figure 3.39, respectively.

Program listing: Figure 3.41 shows the program listing (Program: **RGBfixed**). The program cycles through many colors, where each color is displayed for 2 seconds. The following colors are displayed:

RED, GREEN, BLUE, YELLOW, MAGENTA, WHITE, CYAN

```
//----------------------------------------------------------------------
//                      LED COLOUR WAND - RGB
//                      =====================
```

```
//
// In this project we use an RGB module to generate random colours, just
// like a magic colour wand
//
// Author: Dogan Ibrahim
// File  : RGBfixed
// Date  : June, 2023
//------------------------------------------------------------------------
#define red 2
#define green 3
#define blue 4

//
// Set I/O pin 2, 3 and 4 as outputs
//
void setup()
{
   pinMode(red, OUTPUT);
   pinMode(green, OUTPUT);
   pinMode(blue, OUTPUT);
}

void loop()
{
   //RED
  digitalWrite(red, HIGH);
  digitalWrite(green,LOW);
  digitalWrite(blue,LOW);
  delay(2000);

//GREEN
  digitalWrite(red,LOW);
  digitalWrite(green,HIGH);
  digitalWrite(blue,LOW);
  delay(2000);

//BLUE
  digitalWrite(red,LOW);
  digitalWrite(green,LOW);
  digitalWrite(blue,HIGH);
  delay(2000);

//YELLOW
  digitalWrite(red,HIGH);
  digitalWrite(green,HIGH);
  digitalWrite(blue,LOW);
```

```
  delay(2000);

//MAGENTA
  digitalWrite(red,HIGH);
  digitalWrite(green,LOW);
  digitalWrite(blue,HIGH);
  delay(2000);

//WHITE
  digitalWrite(red,HIGH);
  digitalWrite(green,HIGH);
  digitalWrite(blue,HIGH);
  delay(2000);

//CYAN
  digitalWrite(red,LOW);
  digitalWrite(green,HIGH);
  digitalWrite(blue,HIGH);
  delay(2000);
}
```

Figure 3.41: Program: **RGBfixed**.

3.15 Project 14: Traffic lights

Description: In this project, a program is written to control traffic lights at a simple junction. Red, yellow, and green LEDs are used to simulate the traffic lights. **The kit is supplied with red, blue, and green LEDs. In this project, the yellow LEDs are simulated with the blue LEDs.**

It is assumed that traffic flows from the left-hand side of the road (i.e., the steering wheels of cars are at the right-hand side). In this project, the junction consists of two roads: NORTH ROAD and EAST ROAD. The traffic lights are located at the junction as shown in Figure 3.42.

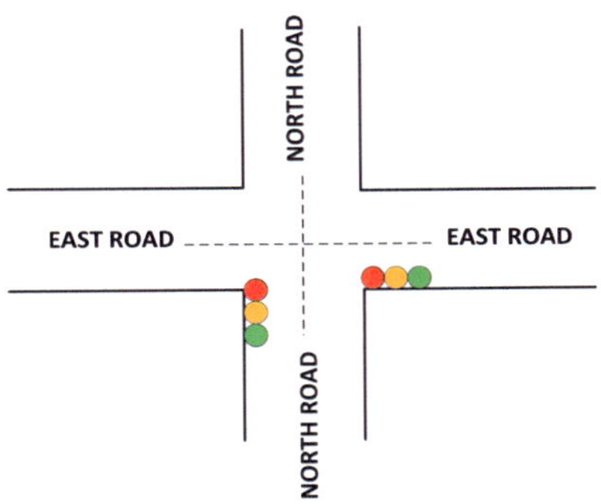

Figure 3.42: Example junction.

The traffic light sequence in the UK is as follows (Figure 3.43). Note that the yellow color is actually amber color:

ROAD 1	**ROAD 2**
RED	GREEN
RED+YELLOW	YELLOW
GREEN	RED
YELLOW	RED+YELLOW
RED	GREEN

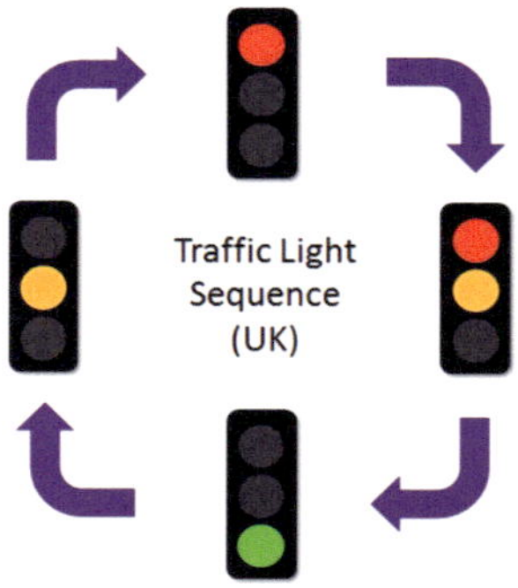

Figure 3.43: Traffic lights sequence in the UK.

The timings used in this project are shown in Table 3.1, where the green on EAST ROAD is for 10 seconds, and the green on NORTH road is for 4 seconds. The cycle time of the lights is 27 seconds, having the following values:

EAST ROAD	**NORTH ROAD**
10 s green	18 s red
12 s red	2 s red+yellow
2 s red+yellow	4 s green
3 s yellow	3 s yellow

Time (seconds)	EAST ROAD	NORTH ROAD
0	GREEN	RED
1	GREEN	RED
2	GREEN	RED
3	GREEN	RED
4	GREEN	RED
5	GREEN	RED
6	GREEN	RED
7	GREEN	RED
8	GREEN	RED
9	GREEN	RED

10	YELLOW	RED
11	YELLOW	RED
12	YELLOW	RED
13	RED	RED
14	RED	RED + AMBER
15	RED	RED + AMBER
16	RED	GREEN
17	RED	GREEN
18	RED	GREEN
19	RED	GREEN
20	RED	YELLOW
21	RED	YELLOW
22	RED	YELLOW
23	RED	RED
24	RED + AMBER	RED
25	RED + AMBER	RED
26	GREEN	RED
27	GREEN	RED

Table 3.1: Timing used in the project.

Block diagram: The block diagram of the project is shown in Figure 3.44, where 6 LEDs are used, 3 for each road (**notice that blue is used to represent "yellow"**).

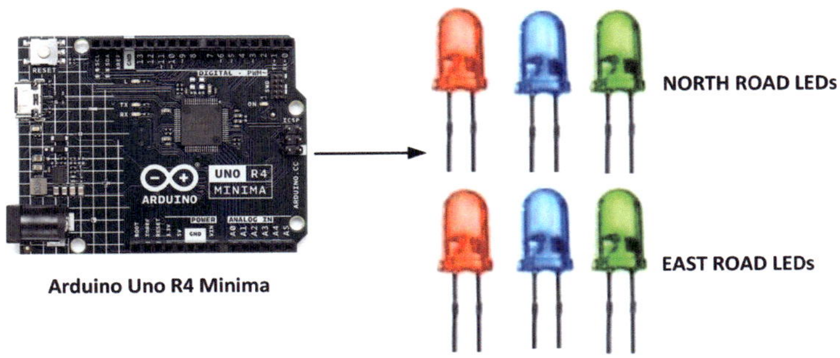

Figure 3.44: Block diagram of the project.

Circuit diagram: Figure 3.45 shows how the LEDs are connected to the development board. NORTH ROAD red, yellow, and green LEDs are connected to port pins 2, 3, and 4, respectively. EAST ROAD red, yellow, and green LEDs are connected to port pins 5, 6, and 7, respectively.

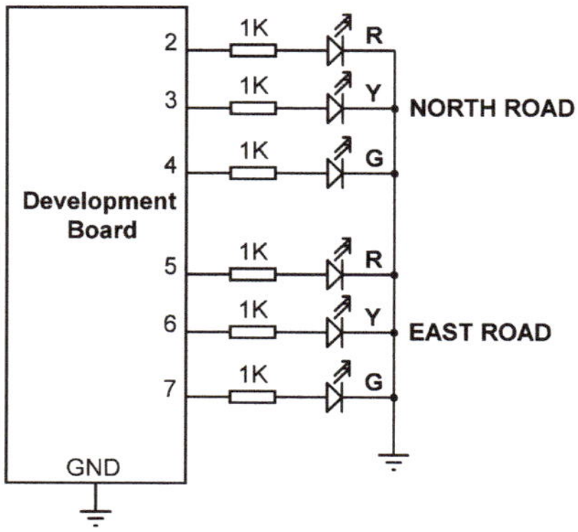

Figure 3.45 Circuit diagram of the project.

Construction: The project was constructed on a breadboard and jumper wires were used to connect the LEDs and resistors to the development board. Figure 3.46 shows the Fritzing component positioning diagram of the project.

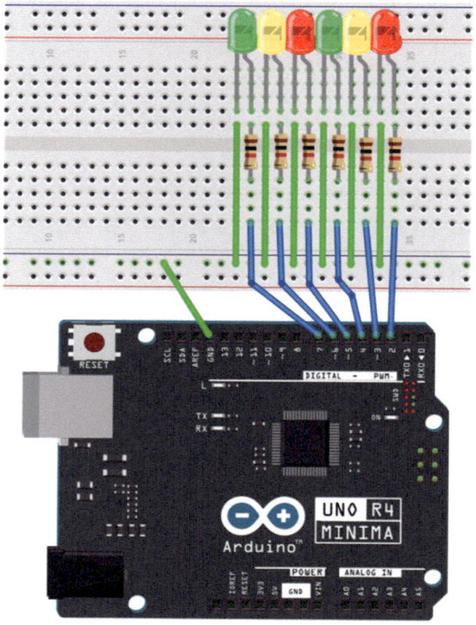

Figure 3.46: Fritzing-style diagram of the project.

Program listing: Figure 3.47 shows the program listing (Program: **traffic**). At the beginning of the program, the LEDs are assigned to output ports and all the LEDs are turned OFF. Function **DelaySeconds()** receives integer **N** as an argument and delays the program by **N** seconds. The main program runs in function **loop()**. Inside this function, the LEDs at both roads are turned ON and OFF at the times specified by Table 3.1.

```
//----------------------------------------------------------------------
//                      TRAFFIC LIGHTS
//                      ==============
//
// This is a traffic lights project. A junction with the road names
// NORTH ROAD and EAST ROAD is considered. The traffic flow is assumed
// to be from the left. The program controls the traffic lights at the
// junction with the timings as described in the text
//
// Author: Dogan Ibrahim
// File  : traffic
// Date  : June, 2023
//----------------------------------------------------------------------
#define NorthR 2                            // NORTH ROAD red
#define NorthY 3                            // NORTH ROAD yellow
#define NorthG 4                            // NORTH ROAD green
#define EastR 5                             // EAST ROAD red
#define EastY 6                             // EAST ROAD yellow
#define EastG 7                             // EAST ROAD green

#define ON HIGH
#define OFF LOW

//
// Set LEDs as outputs and turn them all OFF at the beginning
//
void setup()
{
  for(int i = 0; i < 6; i++)
  {
    pinMode(2+i, OUTPUT);
    digitalWrite(2+i, OFF);
  }
}

//
// N seconds delay
//
void DelaySeconds(int N)
{
  delay(N * 1000);
}

void loop()
{
```

```
  digitalWrite(NorthG, ON);                 // NORTH ROAD green ON
  digitalWrite(EastR, ON);                  // EAST ROAD red ON
  DelaySeconds(10);                         // 10 seconds delay

  digitalWrite(NorthG, OFF);                // NORTH ROAD green OFF
  digitalWrite(NorthY, ON);                 // NORTH ROAD yellow ON
  DelaySeconds(3);                          // 3 seconds delay

  digitalWrite(NorthY, OFF);                // NORTH ROAD yellow OFF
  digitalWrite(NorthR, ON);                 // NORTH ROAD red ON
  DelaySeconds(1);                          // 1 second delay

  digitalWrite(EastY, ON);                  // EAST ROAD red+yellow
  DelaySeconds(2);                          // 2 seconds delay

  digitalWrite(EastR, OFF);                 // EAST ROAD red oFF
  digitalWrite(EastY, OFF);                 // East ROAD yellow OFF
  digitalWrite(EastG, ON);                  // EAST ROAD green ON
  DelaySeconds(4);                          // 4 seconds delay

  digitalWrite(EastG, OFF);                 // EAST ROAD green OFF
  digitalWrite(EastY, ON);                  // EAST ROAD yellow ON
  DelaySeconds(3);                          // 3 seconds delay

  digitalWrite(EastY, OFF);                 // EAST ROAD yellow OFF
  digitalWrite(EastR, ON);                  // EAST ROAD red ON
  DelaySeconds(1);                          // 2 seconds delay

  digitalWrite(NorthY, ON);                 // NOTH ROAD red+yellow ON
  DelaySeconds(2);                          // 2 seconds delay

  digitalWrite(NorthY, OFF);                // NORTHROAD yellow OFF
  digitalWrite(NorthR, OFF);                // NORTH ROAD red OFF
}
```

Figure 3.47: Program: **traffic***.*

3.16 Project 15: Traffic lights with pedestrian crossings

Description: This project is very similar to the previous one but here pedestrian pushbuttons and pedestrian red and green LEDs are used at both the NORTH ROAD and the EAST ROAD. Pressing the pedestrian button stops the traffic on both roads by setting both lights to red for 15 seconds. Access to the pedestrian crossing is only given at the end of a cycle when the lights on both roads are red. If the lights aren't red, then the pedestrian must wait until the cycle is finished and both lights become red. Pressing the pushbutton saves the state of the button and this state is only checked once, when both lights are red. The pedestrian green LED turns ON when it is safe to cross the road, otherwise, the red pedestrian LED is ON.

The traffic lights and the pedestrian crossing buttons are shown in Figure 3.48. Pedestrian crossing pushbuttons are used at one side of each road for simplicity. When the red pedestrian light is ON the pedestrian must wait and only cross the road when the green light is ON.

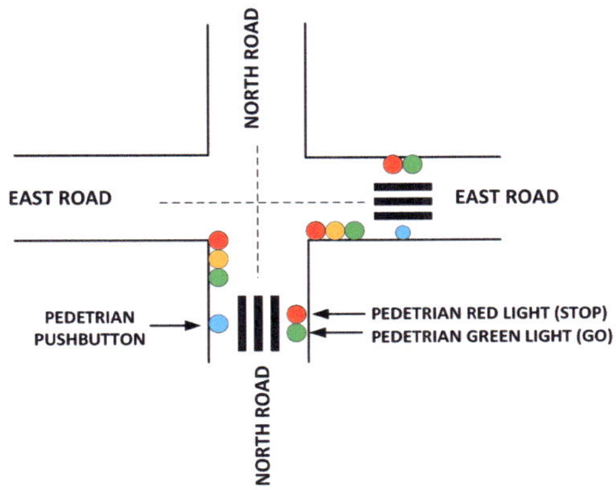

Figure 3.48: Example junction with pedestrian pushbuttons.

Block diagram: The block diagram of the project is shown in Figure 3.49. where 6 LEDs are used, 3 for each road, 2 pushbuttons are used one for each side of the road, and 2 red and 2 green pedestrian lights are used.

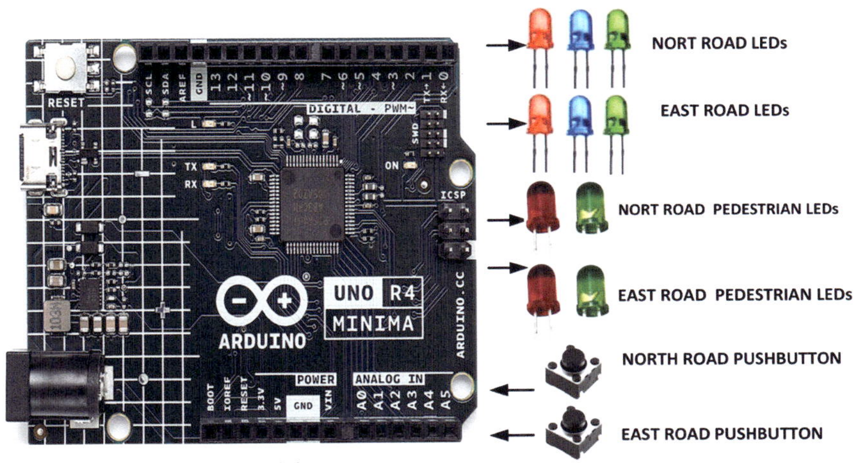

Figure 3.49: Block diagram of the project.

Circuit diagram: The interface between the development board and the LEDs and push-buttons are as follows. Notice that the pushbutton is assigned to port 2 since external interrupts can only be accepted from port 2 or 3:

ROAD EQUIPMENT	PORT NUMBER
NORTH ROAD RED LED	3
NORTH ROAD YELLOW LED	4
NORTH ROAD GREEN LED	5
EAST ROAD RED LED	6
EAST ROAD YELLOW LED	7
EAST ROAD GREEN LED	8
NORTH AND EAST ROAD PEDESTRIAN PUSHBUTTON	2
NORTH AND EAST ROAD PEDESTRIAN RED LED	9
NORTH AND EAST ROAD PEDESTRIAN GREEN LED	10

Figure 3.50 shows the circuit diagram of the project. Notice that the pedestrian LEDs on the NORTH ROAD and EAST ROAD share the same ports. Also, the pedestrian pushbuttons on the NORTH ROAD and EAST ROAD share the same ports.

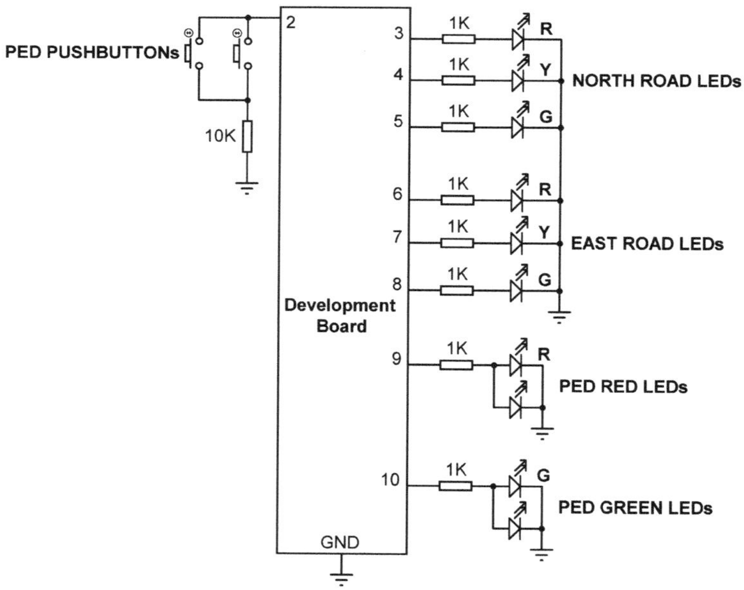

Figure 3.50: Circuit diagram of the project.

Program listing: The program listing is shown in Figure 3.51 (Program**: trafficped**). The program is similar to the one shown in Figure 3.47. Here, additionally, the pedestrian crossing is controlled. At the beginning of the program, all the LEDs and pushbuttons are assigned to the specified ports. The LEDs are configured as outputs and the pushbutton is configured as input with internal pull-up resistors. All the signal LEDs are turned OFF, the red pedestrian LEDs are turned ON, and the green pedestrian LEDs are turned OFF.

The pushbutton is configured as an external interrupt that accepts interrupts on the FALLING edge of the port pin (i.e. when the pushbutton is pressed). On accepting an interrupt, the program jumps to function **PedRequest()** which is the external interrupt service routine. Here, variable **PED** is set to 1 to indicate that the pushbutton has been pressed and there is a pedestrian request. The state of variable **PED** is checked when both roads are in red. If **PED** is set to 1 then the red pedestrian LED is turned OFF and the green pedestrian LED is turned ON to let the pedestrians cross the road. In this project, the pedestrian cycle is set to 15 seconds. At the end of this time, the red pedestrian LED is turned ON and the green pedestrian LED is turned OFF to sign the pedestrians not to cross the road.

```
//-----------------------------------------------------------------------
//               TRAFFIC LIGHTS WITH PEDESTRIAN CROSSINGS
//               ========================================
//
// This is a traffic lights project. A junction with the road names
// NORTH ROAD and EAST ROAD is considered. The traffic flow is assumed
// to be from the left. The program controls the traffic lights at the
// junction with the timings as described in the text.
//
// In this version of the program pedestrian crossings are also controlled
//
// Author: Dogan Ibrahim
// File  : trafficped
// Date  : June, 2023
//-----------------------------------------------------------------------
#define NorthR 3                            // NORTH ROAD red
#define NorthY 4                            // NORTH ROAD yellow
#define NorthG 5                            // NORTH ROAD green
#define EastR 6                             // EAST ROAD red
#define EastY 7                             // EAST ROAD yellow
#define EastG 8                             // EAST ROAD green

#define pedpb 2                             // pedestrian pushbutton
#define pedr 9                              // pedestrian red LED
#define pedg 10                             // pedestrian green LED

#define ON HIGH
#define OFF LOW
int PED = 0;                                // Pedestrian button statue

//
```

```
// Configure the LEDs as outputs and the button as input. ALso,
// configure external interrupts on the pedestrian pushbutton
//
void setup()
{
  for(int i = 0; i < 6; i++)
  {
    pinMode(2+i, OUTPUT);                  // Configure LEDs as outputs
    digitalWrite(2+i, OFF);                // All LEDs except ped LEDs are OFF
    digitalWrite(pedr, ON);                // Pedestrian red LED ON
    digitalWrite(pedg, OFF);               // Pedestrian green LED OFF

    pinMode(pedpb, INPUT_PULLUP);          // pedestrian pushbutton is input

    attachInterrupt(digitalPinToInterrupt(pedpb), PedRequest, FALLING);
  }
}

//
// This is the external interrupt service routine. When the pedestrian pushbutton
// is pressed, the program jumps here to set variable PED to 1 to indicate that
// someone pressed the pushbutton
//
void PedRequest()
{
  PED = 1;
}

//
// N seconds delay
//
void DelaySeconds(int N)
{
  delay(N * 1000);
}

void loop()
{

  digitalWrite(NorthG, ON);                // NORTH ROAD green ON
  digitalWrite(EastR, ON);                 // EAST ROAD red ON
  DelaySeconds(10);                        // 10 seconds delay

  digitalWrite(NorthG, OFF);               // NORTH ROAD green OFF
  digitalWrite(NorthY, ON);                // NORTH ROAD yellow ON
  DelaySeconds(3);                         // 3 seconds delay
```

```
    digitalWrite(NorthY, OFF);                  // NORTH ROAD yellow OFF
    digitalWrite(NorthR, ON);                   // NORTH ROAD red ON
    //
    // At this point the red LEDs are ON at both side of the road
    // Check for pedestrian request
    //
    if(PED == 1)                                // If ped pushbutton pressed
    {
      PED = 0;                                  // Reset ped pushbutton
      digitalWrite(pedg, ON);                   // NORTH ROAD green ped LED ON
      digitalWrite(pedr, OFF);                  // NORTH ROAD red ped LED OFF
      DelaySeconds(15);                         // 15 seconds pedestrian time

      digitalWrite(pedg, OFF);                  // NORTH ROAD green ped LED OFF
      digitalWrite(pedr, ON);                   // NORTH ROAD red ped LED ON
    }
    DelaySeconds(1);                            // 1 second delay

    digitalWrite(EastY, ON);                    // EAST ROAD red+yellow
    DelaySeconds(2);                            // 2 seconds delay

    digitalWrite(EastR, OFF);                   // EAST ROAD red oFF
    digitalWrite(EastY, OFF);                   // East ROAD yellow OFF
    digitalWrite(EastG, ON);                    // EAST ROAD green ON
    DelaySeconds(4);                            // 4 seconds delay

    digitalWrite(EastG, OFF);                   // EAST ROAD green OFF
    digitalWrite(EastY, ON);                    // EAST ROAD yellow ON
    DelaySeconds(3);                            // 3 seconds delay

    digitalWrite(EastY, OFF);                   // EAST ROAD yellow OFF
    digitalWrite(EastR, ON);                    // EAST ROAD red ON
    DelaySeconds(1);                            // 2 seconds delay

    digitalWrite(NorthY, ON);                   // NOTH ROAD red+yellow ON
    DelaySeconds(2);                            // 2 seconds delay

    digitalWrite(NorthY, OFF);                  // NORTHROAD yellow OFF
    digitalWrite(NorthR, OFF);                  // NORTH ROAD red OFF
  }
```

Figure 3.51: Program: **trafficped**.

3.17 Project 16: Using the 74HC595 shift register – binary up counter

The 74HC595 chip

The 74HC595 integrated circuit supplied with the kit is a serial-in parallel-out shift register. Shift registers are used when the number of I/O ports required for a project are not enough and more ports are required. One such example is while controlling 8 LEDs you require 8 ports, but by using a shift register you can connect the LEDs to the parallel output ports of the shift register and send data to the shift register through a clock. This will require only 3 pins (one for the data, one for the serial clock, and one to latch the data), therefore saving 6 I/O pins on your development board. Shift registers become even more important if you want to control say 16 LEDs, which will require 16 ports on the development board.

Data bits are shifted into the LSB of 74HC595 and the bits move to the left at each clock pulse. After 8 clock pulses, 8-bit data will be serially loaded into a shift register. On enabling the Latch pin, the contents of the shift register are copied to its output as a byte. Therefore, the serial data has been converted into parallel form.

Figure 3.52 shows the 74HC595 chip which has the following pins (some manufacturers may use different names for the pins):

Pin	Name	Description
15 and 1 to 7	Q0 to Q7	parallel output
9	Q7S	used to daisy chain other chips
14	DS	serial data input
13	OE	Output enable (LOW to enable)
12	STCP	Latch (Data sent to output when HIGH)
11	SHCP	shift register clock (on the rising edge)
10	MR	Reset (LOW to clear)
8	GND	power supply ground
16	VCC	+5 V

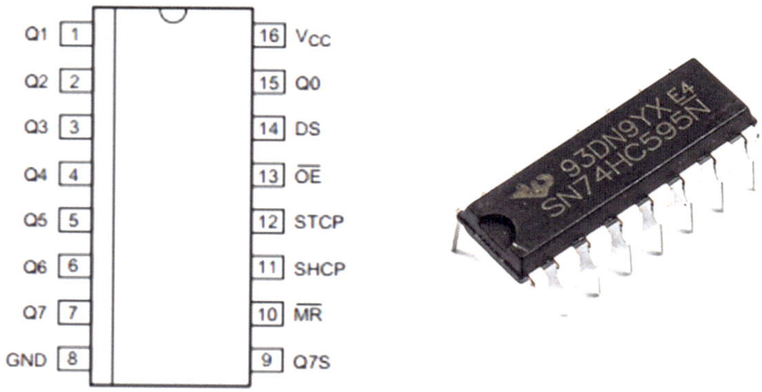

Figure 3.52: The 74HC595 IC.

Description: This is a binary up counter project as in Project 7 (Section 3.8), but here you will be using the 74HC595 IC to control the LEDs.

Block diagram: Figure 3.53 shows the block diagram of the project.

Figure 3.53: Block diagram of the project.

Circuit diagram: Figure 3.54 shows the circuit diagram. Connect VCC and MR pins to +5 V. Also, connect GND and OE pins to power supply GND. Connect the following pins to control the shift register:

Pin 14 (DS) serial data input to pin 4
Pin 12 (STCP) latch to pin 5
Pin 11 (SHCP) clock to pin 6

Now, connect 8 LEDs to pins 1 to 7 and 15 (Q0 to Q7) through 1 kΩ (or smaller value) current-limiting resistors. In this project, the LED at pin 15 is set to be the **M**ost **S**ignificant **B**it (MSB).

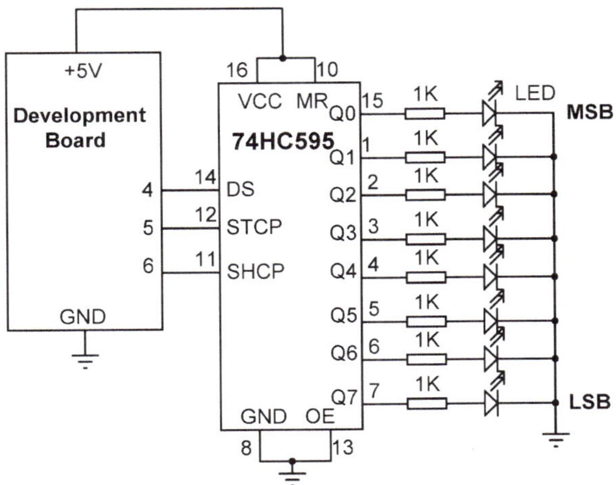

Figure 3.54: Circuit diagram of the project.

Construction: The project was built on a breadboard and connections were made to the development board using jumper wires. Figure 3.55 shows the Fritzing diagram of the project.

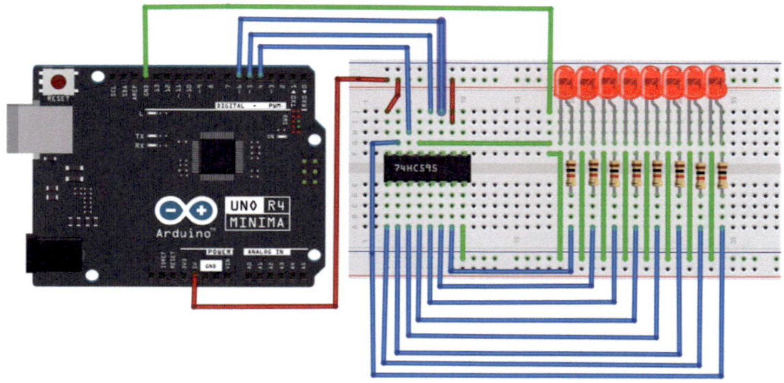

Figure 3.55: Project was built on a breadboard.

Program listing: Figure 3.56 shows the program listing (Program: **ShiftReg**). At the beginning of the program, pins STCP, SHCP, and DS are assigned to 5, 6 and 4 respectively where these are the pins where the chip is connected to the development board. Inside the **setup()** function, these pins are configured as outputs. Inside the main program loop, a **for** loop is executed from 0 to 255 with **N** being the integer variable. **N** is the decimal equivalent of the binary number to be displayed on the LEDs. The latch is set to LOW initially and the data is shifted out of the shift register. By setting the latch high, the data appears at the outputs of the shift register, which then control the LEDs. The loop is repeated after a 500-ms delay.

In this program, the built-in function **shiftOut()** is used. This function has 3 arguments: the data pin, clock pin to be toggled, bit order, and the number to be shifted out in parallel form. The bit order can be **MSBFIRST** or **LSBFIRST**. In this program, **LSBFIRST** is used so that the LED at the LSB position is turned ON first.

```
//----------------------------------------------------------------------
//              74HC595 SHIFT REGISTER LED BINARY COUNTING
//              ==========================================
//
// In this project the 74HC595 shift register chip is used and 8 LEDs
// are connected to the chip.The program counts up in binary from 0 to
// 255 with 500 ms delay between each output
//
// Author: Dogan Ibrahim
// File  : ShiftReg
// Date  : June, 2023
//----------------------------------------------------------------------
int STCP = 5;                                   // Latch pin
int SHCP = 6;                                   // Clock pin
int DS = 4;                                     // Data pin

void setup()
```

```
{
  pinMode(STCP, OUTPUT);                    // Latch is output
  pinMode(SHCP, OUTPUT);                    // Clock is output
  pinMode(DS, OUTPUT);                      // Data is output
}

//
// The main program counts up from 0 to 255 and displays on the LEDs
//
void loop()
{
  for (int N = 0; N < 256; N++)
  {
    digitalWrite(STCP, LOW);                // Latch LOW
    shiftOut(DS, SHCP, LSBFIRST, N);        // Shift out
    digitalWrite(STCP, HIGH);               // Latch HIGH
    delay(500);                             // 500 ms delay
  }
}
```

Figure 3.56: Program: **ShiftReg***.*

3.18 Project 17: Using the 74HC595 shift register — random flashing 8 LEDs

Description: This project is similar to Project 8 (Section 3.9) where 8 LEDs are flashed randomly as if they are Christmas lights. In this project, the LEDs are connected to the 74HC595 shift register.

The block diagram and circuit diagram of the project are in Figure 3.53 and Figure 3.54.

Program listing: Figure 3.57 shows the program listing (Program: **ShiftRandom**). The program is similar to the previous one, but here the built-in function **random()** is used to generate integer random numbers between 1 and 255 and these numbers are sent to the shift register to control the LEDs. The result is that the LEDs flash randomly. 500 ms delay is used between each output.

```
//----------------------------------------------------------------------
//              74HC595 SHIFT REGISTER RANDOM FLASHING LEDs
//              ===========================================
//
// In this project the 74HC595 shift register chip is used and 8 LEDs
// are connected to the chip.The program flashes the LEDs randomly
//
// Author: Dogan Ibrahim
// File  : ShiftRandom
// Date  : June, 2023
```

```
//--------------------------------------------------------------------------
int STCP = 5;                          // Latch pin
int SHCP = 6;                          // Clock pin
int DS = 4;                            // Data pin

void setup()
{
  pinMode(STCP, OUTPUT);               // Latch is output
  pinMode(SHCP, OUTPUT);               // Clock is output
  pinMode(DS, OUTPUT);                 // Data is output
}

//
// The main program counts up from 0 to 255 and displays on the LEDs
//
void loop()
{
  int rnd = random(1, 256);            // Generate random number
  digitalWrite(STCP, LOW);             // Latch LOW
  shiftOut(DS, SHCP, LSBFIRST, rnd);   // Shift out
  digitalWrite(STCP, HIGH);            // Latch HIGH
  delay(500);                          // 500 ms delay
}
```

Figure 3.57: Program: **ShiftRandom***.*

3.19 Project 18: Using the 74HC595 shift register — chasing LEDs

Description: In this project, 8 LEDs are connected to the development board through the 74HC595 shift register. The LEDs chase each other as in Project 5 (Section 3.6), where only one LED is ON at any time. The chasing is from LSB towards MSB with 500 ms between each output.

The block diagram and circuit diagram of the project are in Figure 3.53 and Figure 3.54.

Program listing: Figure 3.58 shows the program listing (Program: **ShiftChase**). The program is very similar to the previous one, but here the number to be sent to the shift register is shifted left at each output.

```
//--------------------------------------------------------------------------
//              74HC595 SHIFT REGISTER CHASING LEDs
//              ===================================
//
// In this project the 74HC595 shift register chip is used and 8 LEDs
// are connected to the chip.The LEDs chase each other
//
```

```
// Author: Dogan Ibrahim
// File  : ShiftChase
// Date  : June, 2023
//----------------------------------------------------------------------
int STCP = 5;                                   // Latch pin
int SHCP = 6;                                   // Clock pin
int DS = 4;                                     // Data pin
int N = 1;

void setup()
{
  pinMode(STCP, OUTPUT);                        // Latch is output
  pinMode(SHCP, OUTPUT);                        // Clock is output
  pinMode(DS, OUTPUT);                          // Data is output
}

//
// The main program counts up from 0 to 255 and displays on the LEDs
//
void loop()
{
  digitalWrite(STCP, LOW);                      // Latch LOW
  shiftOut(DS, SHCP, LSBFIRST, N);              // Shift out
  digitalWrite(STCP, HIGH);                     // Latch HIGH
  delay(500);                                   // 500 ms delay
  N = N << 1;                                   // Shift left
  if(N > 128) N = 1;                            // If MSB lit
}
```

Figure 3.58: Program: **ShiftChase***.*

3.20 Project 19: Using the 74HC595 shift register — turn ON a specified LED

Description: In this project, 8 LEDs are connected to the development board through the shift register as in the previous project. The user is prompted to enter a number between 0 and 7 from the keyboard. The LED corresponding to this number is turned ON for 5 seconds. After this time, all the LEDs are turned OFF and the program repeats. Bit 0 corresponds to the LED at LSB position, and bit 7 corresponds to the LED at the MSB position.

The block diagram and circuit diagram of the project are as in Figure 3.53 and Figure 3.54.

Program listing: Figure 3.59 shows the program listing (Program: **ShiftNo**). A number is read from the keyboard by calling function **Serial.read()**. The user function **SetBit()** is then called. This function receives the bit number and the state to set it to. The built-in function **bitWrite()** is called to set the required bit. This function has three arguments:

the byte whose bit is to be set, the bit number to be set, and the value to set it to (LOW or HIGH). On return from function **SetBit()** the required LED is set for 5 seconds. Notice that function **Serial.read()** reads additional two bytes: the carriage return and line feed. Variable **cr** is used to read these additional bytes in the program.

```
//----------------------------------------------------------------------
//              74HC595 SHIFT REGISTER TURN BIT ON
//              ==================================
//
// In this project the 74HC595 shift register chip is used and 8 LEDs
// are connected to the chip. An integer number is read from the keyboard
// through the Serial Monitor between 0 and 7. The LED corresponsing to
// this number is turned ON where LSB is bit number 0 and MSB is bit
// number 7. After 5 seconds the bit is cleared
//
// Author: Dogan Ibrahim
// File  : ShiftNo
// Date  : June, 2023
//----------------------------------------------------------------------
int STCP = 5;                                // Latch pin
int SHCP = 6;                                // Clock pin
int DS = 4;                                  // Data pin
int N = 1;

void setup()
{
  pinMode(STCP, OUTPUT);                     // Latch is output
  pinMode(SHCP, OUTPUT);                     // Clock is output
  pinMode(DS, OUTPUT);                       // Data is output
  Serial.begin(9600);                        // Serial monitor
  delay(5000);
  Serial.println("Enter the bit no to set: ");
}


//
// This function sets the required bit
//
void SetBit(int Pin, int State)
{
  byte M = 0;
  digitalWrite(STCP, LOW);                   // Latch LOW
  bitWrite(M, Pin, State);                   // Set bit Pin of M
  shiftOut(DS, SHCP, LSBFIRST, M);           // Shift out
  digitalWrite(STCP, HIGH);                  // Latch HIGH
```

```
}
//
// Read the bit number to be set (0 to 7) with LSB bit no 0
//
void loop()
{
  if(Serial.available() > 0)                 // Read a number (0 to 7)
  {
    int N = Serial.read() - '0';
    Serial.println(N);
    SetBit(N, HIGH);                         // Set bit
    delay(5000);                             // Delay 5 seconds
    int cr = Serial.read();                  // Read carriage return
    cr = Serial.read();                      // Read line feed
    SetBit(N, LOW);                          // Bit back to 0
  }
}
```

Figure 3.59: Program: **ShiftNo**.

3.21 Project 20: Using the 74HC595 shift register — turn ON specified LEDs

Description: In this project, 8 LEDs are connected to the development board through the shift register as in the previous project. an integer number is read from the keyboard between 1 and 255. The LED binary pattern corresponding to this number is displayed on the LEDs. For example, if number 17 is entered then the LEDs shown in Figure 3.60 will be ON.

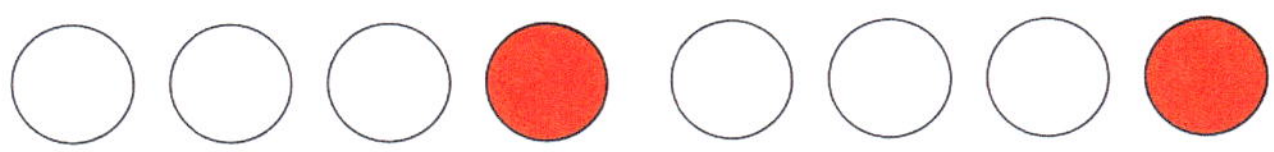

Figure 3.60: LEDs showing binary 17.

The block diagram and circuit diagram of the project are in Figure 3.53 and Figure 3.54.

Program listing: Figure 3.61 shows the program listing (Program: **ShiftDec**). The user is prompted to enter an integer number between 1 and 255. The integer number is read into variable **N** using the built-in function **parseInt()**. Then the LEDs corresponding to this number are turned ON for 3 seconds.

```
//----------------------------------------------------------------------
//              74HC595 SHIFT REGISTER TURN ON SPECIFIED LEDs
//              =============================================
//
// In this project the 74HC595 shift register chip is used and 8 LEDs
// are connected to the chip. An integer number is read from the keyboard
// through the Serial Monitor between 1 and 255. The LEDs corresponsing to
```

```
// this number are turned ON. After 3 seconds the LEDs are cleared
//
// Author: Dogan Ibrahim
// File  : ShiftDec
// Date  : June, 2023
//-------------------------------------------------------------------------
int STCP = 5;                               // Latch pin
int SHCP = 6;                               // Clock pin
int DS = 4;                                 // Data pin

void setup()
{
  pinMode(STCP, OUTPUT);                    // Latch is output
  pinMode(SHCP, OUTPUT);                    // Clock is output
  pinMode(DS, OUTPUT);                      // Data is output
  Serial.begin(9600);                       // Serial monitor
  delay(5000);
  Serial.println("Enter a number 1 - 255: ");
}


//
// Read the bit number to be set (0 to 7) with LSB bit no 0
//
void loop()
{
  if(Serial.available() > 0)                // Read a number (0 to 7)
  {
    int N = Serial.parseInt();              // Get integer number
    Serial.println(N);
    int cr = Serial.read();                 // Carriage return
    digitalWrite(STCP, LOW);                // Latch LOW
    shiftOut(DS, SHCP, LSBFIRST, N);        // Shift out
    digitalWrite(STCP, HIGH);               // Latch HI
    delay(3000);                            // Delay 3 secs
  }
}
```

Figure 3.61: Program: ***ShiftDec.***

Chapter 4 • 7-Segment LED Displays

4.1 Overview

In the preceding chapter, you have learned how to develop projects using LEDs only. In this chapter, you will be using 7-segment LED displays in projects. Two types of 7-segment LEDs are supplied in the kit: 1-digit, and 4-digit.

4.2 7-Segment LED display structure

7-segment LED displays are used frequently in electronic circuits to show numeric or alphanumeric values. As shown in Figure 4.1, a 7-segment LED display basically consists of 7 LEDs connected such that numbers from 0 to 9 and some letters can be displayed. Segments are identified by letters from **a** through **g** and Figure 4.2 shows the segment names of a typical 7-segment LED display.

Figure 4.1: Some 7-segment LED displays.

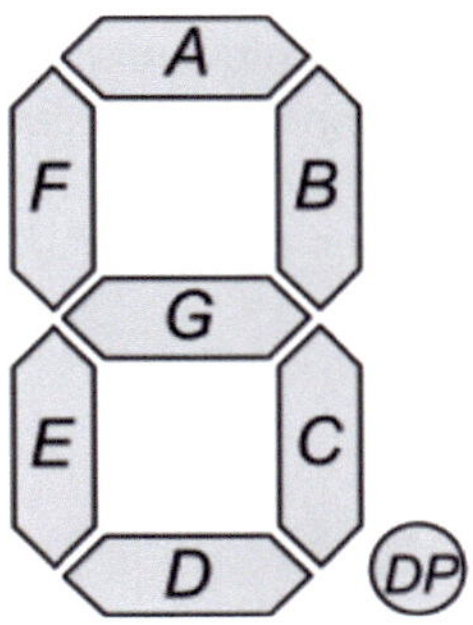

Figure 4.2: Segment names of a 7-segment display.

Figure 4.3 shows how numbers from 0 to 9 can be obtained by turning ON different segments of the display.

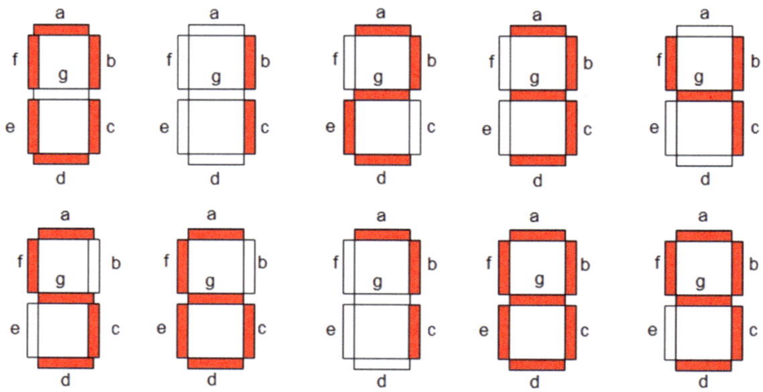

Figure 4.3: Displaying numbers 0–9 (courtesy of electronics-fun.com).

7-segment displays are available in two different configurations: **common-cathode** and **common-anode**. As shown in Figure 4.4, in common-cathode configuration all the cathodes of all segment LEDs are connected together to ground or the lowest potential. The segments are turned ON by applying a logic 1 to the required segment LED via current-limiting resistors. In common-cathode configuration, the 7-segment LED is connected to the microcontroller in current sourcing mode.

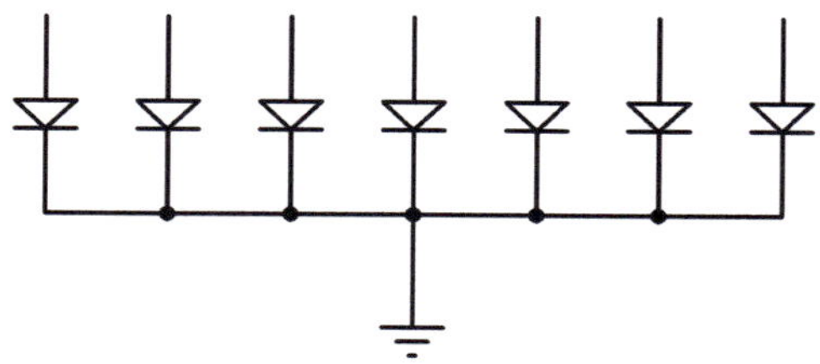

Figure 4.4: Common-cathode 7-segment display.

In a common-anode configuration, the anode terminals of all the LEDs are connected together as shown in Figure 4.5. This common point is then normally connected to the supply voltage. A segment is turned ON by connecting its cathode terminal to logic 0 via a current-limiting resistor. In common-anode configuration, the 7-segment LED is connected to the microcontroller in current sinking mode.

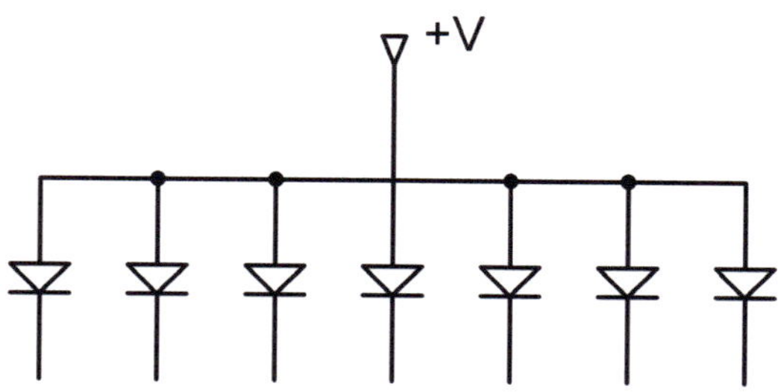

Figure 4.5: Common-cathode 7-segment display.

4.3 Project 1: 7-Segment 1-digit LED counter

Description: In this project, you will use the 1-digit display contained in the kit to count from 0 to 9.

Block diagram: Figure 4.6 shows the project block diagram.

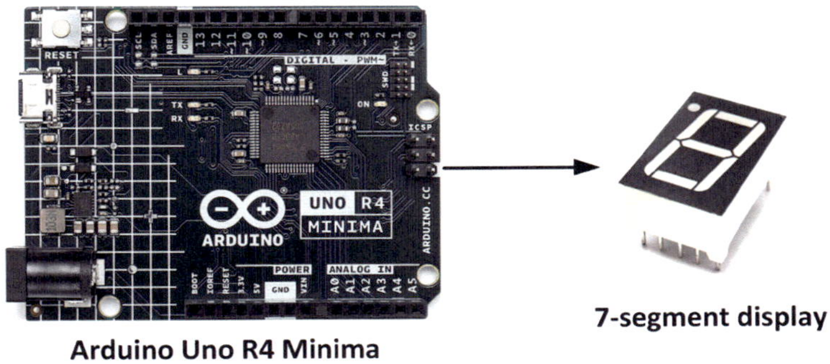

Figure 4.6: Block diagram of the project.

Circuit diagram: The 7-segment display supplied with the kit is type 5161AH, red, common-cathode type with a digit height of 0.56 inches. Figure 4.7 shows the display structure with its connection diagram. This is a 10-pin display with pin 1 located as shown in the image. Pin numbering is 1 through 5 going right from pin 1. Pins 6 through 10 are located at the top part of the display with pin 6 at the top right-hand side. Notice where pin 1 is situated.

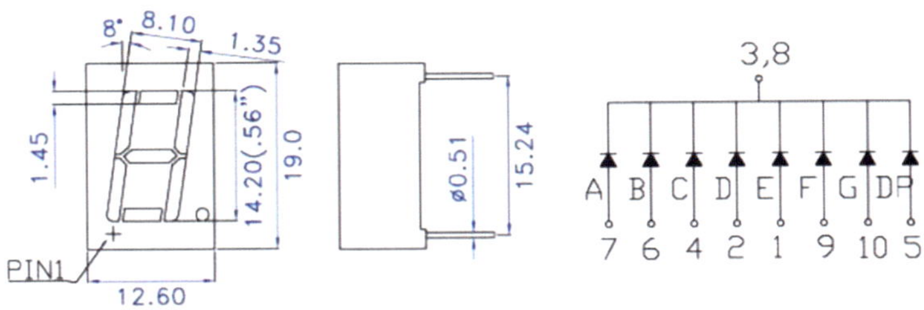

Figure 4.7: The supplied display.

Figure 4.8 shows the display connected to the development board through current-limiter resistors. The common pins of the display are connected to GND.

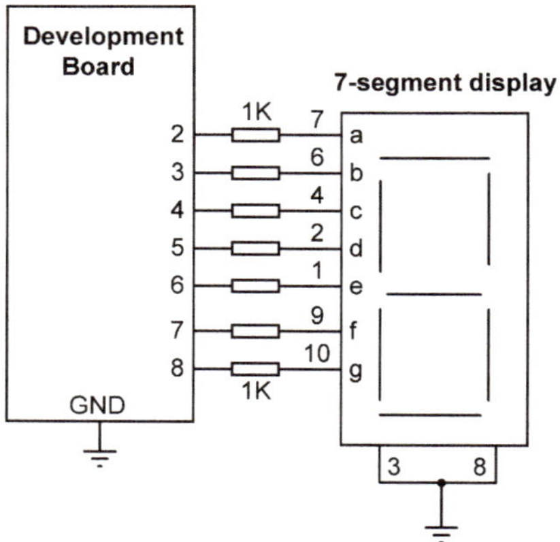

Figure 4.8: Circuit diagram of the project.

The connections between the ports and the 7-segment LED pins are as follows:

7-segment LED pin	**Port pin**
a (pin 7)	2
b (pin 6)	3
c (pin 4)	4
d (pin 2)	5
e (pin 1)	6
f (pin 9)	7
g (pin 10)	8
common (pins 3,8)	GND

Before using a 7-segment display, you have to map the pins to the MCU ports. Table 4.1 shows the number to be displayed and the data to be sent to the port pins. Notice that the MSB bit (shown as x in the table) is not used and is set to 0.

Number	x g f e d c b a	Port data in Hex
0	0 0 1 1 1 1 1 1	0x3F
1	0 0 0 0 0 1 1 0	0x06
2	0 1 0 1 1 0 1 1	0x5B
3	0 1 0 0 1 1 1 1	0x4F
4	0 1 1 0 0 1 1 0	0x66
5	0 1 1 0 1 1 0 1	0x6D
6	0 1 1 1 1 1 0 1	0x7D
7	0 0 0 0 0 1 1 1	0x07
8	0 1 1 1 1 1 1 1	0x7F
9	0 1 1 0 1 1 1 1	0x6F

Table 4.1: Number to be displayed and port data.

Program listing: Figure 4.9 shows the program listing (Program: **SevenSeg1**). Array **LED** stores the pin numbers corresponding to the connections between the display and the development board ports. Array **SEG** stores the mapping between the numbers and the data to be sent to display a number as shown in Table 4.1. At the beginning of the program, the used ports are configured as outputs. Function **Display()** groups a number of port pins and sends data to the group. This function has two arguments: the number to be displayed, and the data width. Here, 8 port pins are grouped together (MSB bit is not used) and hexadecimal byte data is sent to the group as shown in Table 4.1. Variable **Count** stores the number to be sent to the display. The program displays numbers 0 to 9 continuously with one second delay between each display. Notice that **Count** is reset to 0 when it reaches 10.

```
//----------------------------------------------------------------------
//                        7-SEGMENT DISPLAY COUNTER
//                        =========================
//
// This is a 7-segment LED counter which counts up every second 0 to 9
//
// Author: Dogan Ibrahim
// File  : SevenSeg1
// Date  : June, 2023
//----------------------------------------------------------------------
#define ON HIGH                                     // Define ON
#define OFF LOW                                     // Define OFF
int LED[] = {9, 8, 7, 6, 5, 4, 3, 2};               // LEDs at ports 2 to 8
unsigned char SEG[] = {0x3F,0x06,0x5B,0x4F,0x66,    // See Table 4.1
              0x6D, 0x7D, 0x07, 0x7F, 0x6F};

int Count = 0;                                      // COunt=0 to start with

void setup()
{
  for(int i = 0; i < 8; i++)
  {
    pinMode(LED[i], OUTPUT);                        // Configure as outputs
  }
}


//
// Group the port pins together. L is the number of bits (8 here),and No
// is the data to be displayed
//
void Display(int No, int L)
{
  int i, m, j;
```

```
  m = L - 1;
  for(i = 0; i < L; i++)
  {
    j = 1;
    for(int k = 0; k < m; k++)j = j * 2;
    if((No & j) != 0)
      digitalWrite(LED[i], ON);
    else
      digitalWrite(LED[i], OFF);
    m--;
  }
}


void loop()
{
  unsigned char Pattern = SEG[Count];          // Get number to send to
  Display(Pattern, 8);                         // Display the number
  Count++;                                     // Increment Count
  if(Count == 10) Count = 0;                   // If Count=10, set to 0
  delay(1000);                                 // 1 second delay
}
```

Figure 4.9: Program: **SevenSeg1**.

Suggestion
Note that the program can be made more readable if you create a function to display the required number and then call this function from the main program.

4.4 Project 2: 7-Segment 4-digit multiplexed LED display

Project Description
This project is similar to the previous project but here multiplexed four digits are used instead of one digit. The program displays fixed number '5346' on the 7-segment display.

Block diagram: Figure 4.10 shows the project block diagram.

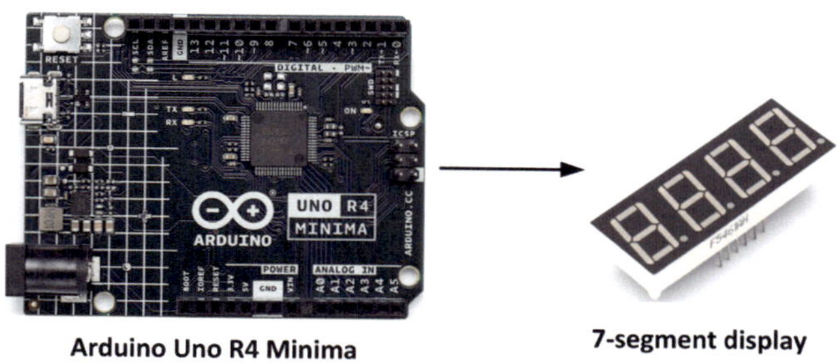

Figure 4.10: Block diagram of the project.

Circuit diagram: The 7-segment 4-digit display supplied with the kit is a red, common-cathode type, with digit heights of 0.36 inches. Figure 4.11 shows the display structure with its connection diagram. The display features one decimal point per digit. The LEDs have a forward voltage of 1.8 VDC and a max. forward current of 30 mA. The hardware interface is 12 (two rows of 6) through-hole pins. Pin 1 is located as shown in the image. Pins 1 to 6 are at the bottom, and pins 7 to 12 are at the top with pin 7 at the top right-hand side. Corresponding a-g pins of each digit are connected together. Digit 1 is the leftmost digit.

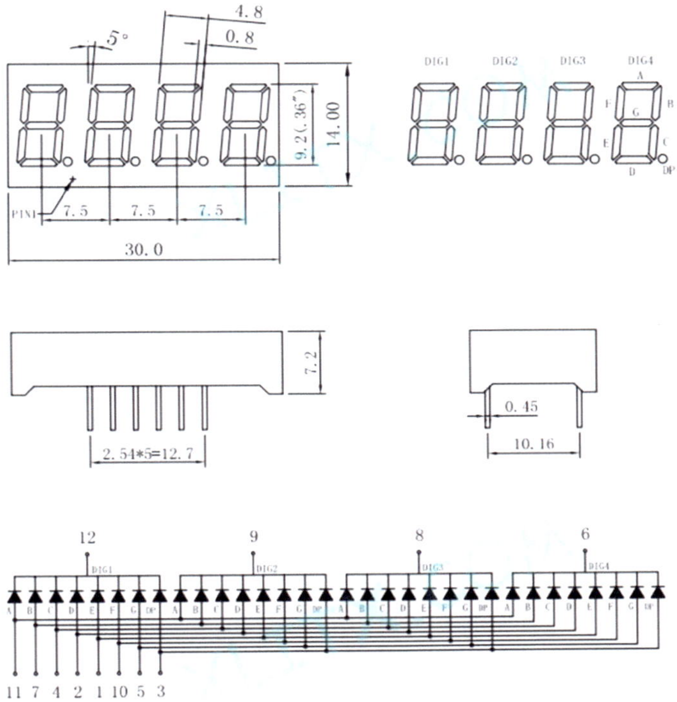

Figure 4.11: The supplied display.

The individual digits of multiplexed 7-segment displays are normally enabled using transistors as shown in Figure 4.12. This is because the current sourcing/sinking capabilities of the MCU may not be enough to turn ON the required segments. In general, any NPN-type transistor can be used (e.g., the BC337).

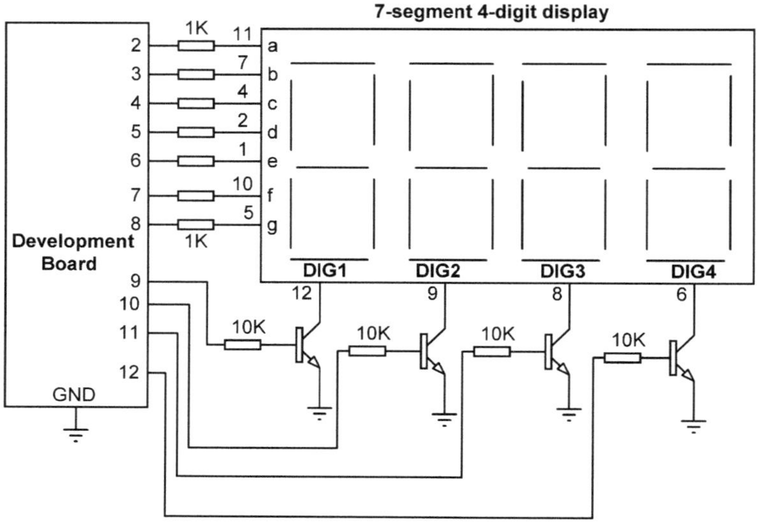

Figure 4.12: Enabling digits using transistors.

In Arduino Uno R4-based projects, each I/O pin can handle 8 mA of current and in total no more than 60 mA should be drawn from all the GPIO ports. With 1-kohm current-limiting resistors, each segment draws about 3 mA. If all segments of a digit are ON (e.g., displaying number 8), the required current for the digit is 3 × 7 = 21 mA which is too high for the Uno R4 GPIO pins. It is therefore necessary to use transistor switch circuits (Figure 4.13).

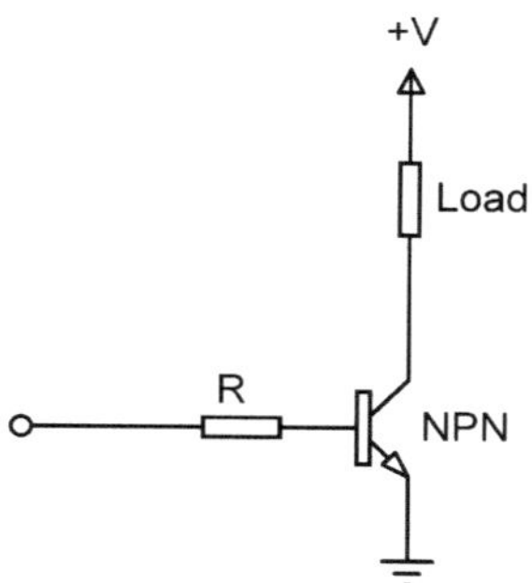

Figure 4.13: Transistor switch circuit.

The connections between the ports and the 7-segment LED pins are as follows:

7-segment	LED pin	Port pin
a	(pin 11)	2
b	(pin 7)	3
c	(pin 4)	4
d	(pin 2)	5
e	(pin 1)	6
f	(pin 10)	7

g	(pin 5)	8
DIG1	(pin 12)	9 via transistor
DIG2	(pin 9)	10 via transistor
DIG3	(pin 8)	11 via transistor
DIG4	(pin 6)	12 via transistor

Program Listing: By displaying each digit for several milliseconds the eye can not differentiate that the digits are not ON all the time. This way you can multiplex any number of 7-segment displays together. For example, to display the number 5346, you have to send 5 to the first digit and enable its common pin. After a few milliseconds, number 3 is sent to the second digit and the common point of the second digit is enabled, and so on When this process is repeated continuously, the user sees as if both displays are ON continuously.

Figure 4.14 shows the program listing (Program: **SevenSeg2**). At the beginning of the program, the display pins are assigned to the ports. Array LED[] stores the LED segment port assignments, array SEG[] stores the bit pattern to turn on a digit, and array DIGITS[] stores the digit port assignments. Inside the **setup()** function, all LED pins and digit pins are configured as output and the digits are disabled. The program is similar to the one with one digit. Here, number 5 is sent to digit 1 (MSD) and **DIGITS[0]** is enabled for about 5 milliseconds. Then, **DIGITS[0]** is disabled and number 3 is sent to digit 2 and **DIGITS[1]** is enabled for about 5 milliseconds. Then, **DIGITS[1]** is disabled and number 4 is sent to the third digit and **DIGITS[2]** is enabled for about 5 milliseconds. Finally, **DIGITS[2]** is disabled and number 6 is sent to the last digit (LSD) and **DIGITS[3]** is enabled for about 5 milliseconds. This process is repeated after disabling **DIGITS[3]**, thus displaying number '5346' on the 7-segment display.

```
//----------------------------------------------------------------------
//                      7-SEGMENT 4-DIGIT DISPLAY
//                      =========================
//
// This is a 7-segment 4-digit display program. Number 5346 is displayed
//
// Author: Dogan Ibrahim
// File  : SevenSeg2
// Date  : June, 2023
//----------------------------------------------------------------------
#define ON HIGH                                     // Define ON
#define OFF LOW                                     // Define OFF
int LED[] = {8, 7, 6, 5, 4, 3, 2};                  // LEDs at ports 2 to 8
unsigned char SEG[] = {0x3F,0x06,0x5B,0x4F,0x66,    // See Table 4.1
             0x6D, 0x7D, 0x07, 0x7F, 0x6F};
unsigned char DIGITS[] = {9, 10, 11, 12};           // DIGIT ports
unsigned int Cnt = 5346;                            // Number to display
unsigned int Dig1, Dig2, Dig3, Dig4, Dig5, Dig6;
int Pattern;
#define Enable HIGH
```

```
#define Disable LOW

//
// COnfigure LED segments and digits as outputs
//
void setup()
{
  for(int i = 0; i < 7; i++)
  {
    pinMode(LED[i], OUTPUT);                    // Configure as outputs
  }

  for(int i = 0; i < 4; i++)
  {
    pinMode(DIGITS[i], OUTPUT);                 // Configure as outputs
    digitalWrite(DIGITS[i], Disable);           // All digits are OFF
  }
}


//
// Group the port pins together. L is the number of bits (8 here),and No
// is the data to be displayed
//
void Display(int No, int L)
{
  int i, m, j;

  m = L - 1;
  for(i = 0; i < L; i++)
  {
    j = 1;
    for(int k = 0; k < m; k++)j = j * 2;
    if((No & j) != 0)
      digitalWrite(LED[i], ON);
    else
      digitalWrite(LED[i], OFF);
    m--;
  }
}


void loop()
{
  Dig1 = Cnt / 1000;                            // 1000s  digit
  Pattern = SEG[Dig1];                          // Get the bit pattern
```

```
  Display(Pattern, 7);                                // Send to display
  digitalWrite(DIGITS[0], Enable);                    // Enable DIG1
  delay(5);                                           // Wait a while
  digitalWrite(DIGITS[0], Disable);                   // Disable DIG1

  Dig2 = Cnt % 1000;
  Dig3 = Dig2 / 100;                                  // 100s digit
  Pattern =  SEG[Dig3];                               // Get the bit pattern
  Display(Pattern, 7);                                // Send to display
  digitalWrite(DIGITS[1], Enable);                    // Enable DIG2
  delay(5);                                           // Wait a while
  digitalWrite(DIGITS[1], Disable);                   // Disable DiG2

  Dig4 = Dig2 % 100;
  Dig5 = Dig4/10;                                     // 10s digit
  Pattern = SEG[Dig5];                                // Get the bit pattern
  Display(Pattern, 7);                                // Send to display
  digitalWrite(DIGITS[2], Enable);                    // Enable DIG3
  delay(5);                                           // Wait a while
  digitalWrite(DIGITS[2], Disable);                   // Disable DIG3

  Dig6 = Dig4 % 10;                                   // 1s digit
  Pattern = SEG[Dig6];                                // Get the bit pattern
  Display(Pattern, 7);                                // Send to display
  digitalWrite(DIGITS[3], Enable);                    // Enable DIG4
  delay(5);                                           // Wait a while
  digitalWrite(DIGITS[3], Disable);                   // Disable DIG4
}
```

Figure 4.14: Program: **SevenSeg2**.

4.5 Project 3: 7-Segment 4-digit multiplexed LED display counter – timer interrupts

Why timer interrupts?

In the previous example, you displayed a fixed number on the 7-segment display. This was not a real application since you may almost always want to display different numbers. For example, during counting different numbers are displayed. The problem here is that the 7-segment display has to be refreshed nearly every few milliseconds and the CPU cannot refresh the display and at the same time execute other user code as it requires multitasking. The solution is to refresh the display in a timer interrupt routine and carry out the normal user tasks in the main program.

In this program, you will refresh the display in the timer interrupt service routine and then send data to the display in the main program. Before doing this, it is worthwhile to review the Arduino UNO R4 timers and timer interrupts.

Timer interrupts

The Arduino UNO R4 is based on the Renesas RA4M1 processor. This processor has the following built-in timers:

2× 32-bit General PWM timer (GPT32)
6× 16-bit General PWM timer (GPT16)
2× 16-bit Asynchronous General Purpose timer (AGT)

Watchdog timer (WDT)

The General PWM Timer (GPT) is a 32-bit timer with 2 channels and a 16-bit timer with 6 channels. PWM waveforms can be generated by controlling the up-counter, down-counter, or the up- and down-counter. In addition, PWM waveforms can be generated for controlling brushless DC motors. The GPT can also be used as a general-purpose timer. You can explicitly request a timer that has been reserved for PWM.

The Asynchronous General Purpose Timer (AGT) is a 16-bit timer that can be used for pulse output, external pulse width or period measurement, and counting of external events. One of these timers is used to provide Arduino millis() and microseconds() methods. This 16-bit timer consists of a reload register and a down counter. The reload register and the down counter are allocated to the same address, and they can be accessed with the AGT register.

The Watchdog Timer (WDT) is a 14-bit down-counter. It can be used to reset the MCU when the counter underflows because the system has run out of control and is unable to refresh the WDT. In addition, a non-maskable interrupt or interrupt can be generated by an underflow. A refresh-permitted period can be set to refresh the counter and used as the condition to detect when the system runs out of control.

Description: In this project, you will count up every second and display on the 7-segment display. A timer will be used to refresh the display every 5 milliseconds inside the timer interrupt service routine.

The block diagram and circuit diagram of the project are in Figure 4.10 and 4.13 respectively.

Program listing: Figure 4.15 shows the program listing (Program: **SevenSeg3**). The program uses the Arduino Uno R4 FspTimer core functions. At the beginning of the program, the **FspTimer** header is included and the interface between the 7-segment multiplexed display and the development board I/O ports are defined and all set as outputs. All the digits are disabled inside the **setup()** function. Function **Display()** groups the I/O bits together as a port so that the bits can be accessed together. Function **StartTimer()** gets an available GPT timer with its index set to **TimerIndex**. The timer mode is set to periodic, its frequency is set to **freq (200)**, and the interrupt service routine is named the function **TimerCallback()**. The allocated timer is then opened and started. Inside the **TimerCallback()** function, the digits are sent to the 7-segment display with each digit being displayed for 5 milliseconds. i.e., the display is refreshed every 5 milliseconds. Variable **flag** determines which digit should be refreshed such that if flag = 0 the digit 1 is refreshed and so on. Inside the main program loop, variable **cnt** is incremented by one. When cnt is greater than 9999, then it is reset to 0 and counting continues. A one-second delay is used between each output count.

```
//-----------------------------------------------------------------------
//                    7-SEGMENT 4-DIGIT DISPLAY COUNTER
//                    =================================
//
// This is a 7-segment 4-digit display program. The display counts up
// every second. Timer interrupts are used to refresh the display. The
// timer is configured to interrupt at every 5 ms (i.e 200 Hz)
//
// Author: Dogan Ibrahim
// File  : SevenSeg3
// Date  : June, 2023
//-----------------------------------------------------------------------
#include "FspTimer.h"
FspTimer MyTimer;

#define ON HIGH                                        // Define ON
#define OFF LOW                                        // Define OFF
int LED[] = {8, 7, 6, 5, 4, 3, 2};                     // LEDs at ports 2 to 8
unsigned char SEG[] = {0x3F,0x06,0x5B,0x4F,0x66,       // See Table 4.1
0x6D, 0x7D, 0x07, 0x7F, 0x6F};
unsigned char DIGITS[] = {9, 10, 11, 12};              // DIGIT ports
unsigned int MSD, m, MID2, n, MID1, LSD;
unsigned char Pattern;
int cnt = 0;
int flag = 0;
uint8_t TimerType;
int8_t TimerIndex;
#define Enable HIGH
#define Disable LOW

//
// Group the port pins together. L is the number of bits (8 here),and No
// is the data to be displayed
//
void Display(int No, int L)
{
  int i, m, j;

  m = L - 1;
  for(i = 0; i < L; i++)
  {
    j = 1;
```

```
    for(int k = 0; k < m; k++)j = j * 2;
    if((No & j) != 0)
      digitalWrite(LED[i], ON);
    else
      digitalWrite(LED[i], OFF);
    m--;
  }
}

//
// TIMER interrupt service routine (every 5 ms, 200 Hz)
//
void TimerCallback(timer_callback_args_t __attribute((unused)) *p_args)
{
  MSD = cnt / 1000;                                  // Get MSD
  m = cnt % 1000;
  MID2 = m / 100;                                    // Get MID2
  n = m % 100;
  MID1 = n / 10;                                     // Get MID1
  LSD = n % 10;                                      // Get LSD

  if(flag == 0)
  {
    digitalWrite(DIGITS[3], Disable);                // Disable DIG4
    Pattern = SEG[MSD];                              // Get pattern
    Display(Pattern, 7);                             // Display number
    digitalWrite(DIGITS[0], Enable);                 // Enable DIG1
    flag++;
  }
  else if(flag == 1)
  {
    digitalWrite(DIGITS[0], Disable);                // Disable DIG1
    Pattern = SEG[MID2];                             // Get pattern
    Display(Pattern, 7);                             // Display number
    digitalWrite(DIGITS[1], Enable);                 // Enable DIG2
    flag++;
  }
  else if(flag == 2)
  {
    digitalWrite(DIGITS[1], Disable);                // DIsable DIG2;
    Pattern = SEG[MID1];                             // Get pattern
    Display(Pattern, 7);                             // Display number
    digitalWrite(DIGITS[2], Enable);                 // Enable DIG3
    flag++;
  }
  else if(flag == 3)
```

```
    {
      digitalWrite(DIGITS[2], Disable);             // Disable DIG3
      Pattern = SEG[LSD];                           // Get pattern
      Display(Pattern, 7);                          // Display number
      digitalWrite(DIGITS[3], Enable);              // Enable DIG4
      flag = 0;
    }
}

//
// Get a GPT timer, set its mode, define callback ISR and start the timer
//

void StartTimer(float freq)
{
  TimerType = GPT_TIMER;
  TimerIndex = FspTimer::get_available_timer(TimerType);
  if (TimerIndex == 0)
  {
    FspTimer::force_use_of_pwm_reserved_timer();
    TimerIndex = FspTimer::get_available_timer(TimerType);
  }

  MyTimer.begin(TIMER_MODE_PERIODIC, TimerType, TimerIndex, freq, 0.0f,
TimerCallback);
  MyTimer.setup_overflow_irq();
  MyTimer.open();
  MyTimer.start();
}

//
// Configure segment LEDs as outputs, set timer for 200 Hz
//
void setup()
{
  for(int i = 0; i < 7; i++)
  {
    pinMode(LED[i], OUTPUT);                        // Configure as outputs
  }

  for(int i = 0; i < 4; i++)
  {
    pinMode(DIGITS[i], OUTPUT);                     // Configure as outputs
    digitalWrite(DIGITS[i], Disable);               // All digits are OFF
  }
```

```
//
// Timer1 configuration for 5ms (200Hz) interrupts
//
  StartTimer(200);                              // Set for 200 Hz
}

//
// Main program loop. Increment count here and delay 1 second
//
void loop()
{
  cnt++;                                      // Increment count
  if(cnt > 9999) cnt = 0;                     // If 9999, reset to 0
  delay(1000);                                // Delay 1 second
}
```

Figure 4.15: Program: ***SevenSeg3****.*

4.6 Project 4: 7-Segment 4-digit multiplexed LED display counter — blanking leading zeroes

Description: In the program in Figure 4.15 the leading digits are shown as 0 when the number is lower than these digits. For example, the number 12 is displayed as 0012 and not as 12. That's not desirable in many applications. In this project, let's disable the leading zeroes.

Program listing: Figure 4.16 shows the modified program (Program: **SevenSeg4**) shows how you can blank the leading 0s by disabling their digits. For example, digit 1 is enabled if the number to be displayed is greater than 999, digit 2 is enabled if the number is greater than 99 and so on.

```
//-----------------------------------------------------------------------
//                    7-SEGMENT 4-DIGIT DISPLAY COUNTER
//                    =================================
//
// This is a 7-segment 4-digit display program. The display counts up
// every second. Timer interrupts are used to refresh the display. The
// timer is configured to interrupt at every 5 ms (i.e 200 Hz)
//
// In this program the leading zeroes are disabled so that for example
// number 12 is shown as   12 and not as 0012
//
// Author: Dogan Ibrahim
// File  : SevenSeg4
// Date  : June, 2023
//-----------------------------------------------------------------------
#include "FspTimer.h"
```

```
FspTimer MyTimer;

#define ON HIGH                                         // Define ON
#define OFF LOW                                         // Define OFF
int LED[] = {8, 7, 6, 5, 4, 3, 2};                      // LEDs at ports 2 to 8
unsigned char SEG[] = {0x3F,0x06,0x5B,0x4F,0x66,        // See Table 4.1
0x6D, 0x7D, 0x07, 0x7F, 0x6F};
unsigned char DIGITS[] = {9, 10, 11, 12};               // DIGIT ports
unsigned int MSD, m, MID2, n, MID1, LSD;
unsigned char Pattern;
int cnt = 0;
int flag = 0;
uint8_t TimerType;
int8_t TimerIndex;
#define Enable HIGH
#define Disable LOW

//
// Group the port pins together. L is the number of bits (8 here),and No
// is the data to be displayed
//
void Display(int No, int L)
{
  int i, m, j;

  m = L - 1;
  for(i = 0; i < L; i++)
  {
    j = 1;
    for(int k = 0; k < m; k++)j = j * 2;
    if((No & j) != 0)
      digitalWrite(LED[i], ON);
    else
      digitalWrite(LED[i], OFF);
    m--;
  }
}

//
// TIMER interrupt service routine (every 5 ms, 200 Hz)
//
void TimerCallback(timer_callback_args_t __attribute((unused)) *p_args)
{
  MSD = cnt / 1000;                                     // Get MSD
  m = cnt % 1000;
  MID2 = m / 100;                                       // Get MID2
```

```
    n = m % 100;
    MID1 = n / 10;                                           // Get MID1
    LSD = n % 10;                                            // Get LSD

    if(flag == 0)
    {
      digitalWrite(DIGITS[3], Disable);                      // Disable DIG4
      Pattern = SEG[MSD];                                    // Get pattern
      Display(Pattern, 7);                                   // Display number
      if(cnt > 999)digitalWrite(DIGITS[0], Enable); // Enable DIG1
      flag++;
    }
    else if(flag == 1)
    {
      digitalWrite(DIGITS[0], Disable);                      // Disable DIG1
      Pattern = SEG[MID2];                                   // Get pattern
      Display(Pattern, 7);                                   // Display number
      if(cnt > 99)digitalWrite(DIGITS[1], Enable);  // Enable DIG2
      flag++;
    }
    else if(flag == 2)
    {
      digitalWrite(DIGITS[1], Disable);                      // DIsable DIG2;
      Pattern = SEG[MID1];                                   // Get pattern
      Display(Pattern, 7);                                   // Display number
      if(cnt > 9)digitalWrite(DIGITS[2], Enable);   // Enable DIG3
      flag++;
    }
    else if(flag == 3)
    {
      digitalWrite(DIGITS[2], Disable);                      // Disable DIG3
      Pattern = SEG[LSD];                                    // Get pattern
      Display(Pattern, 7);                                   // Display number
      digitalWrite(DIGITS[3], Enable);                       // Enable DIG4
      flag = 0;
    }
}

//
// Get a GPT timer, set its mode, define callback ISR and start the timer
//

void StartTimer(float freq)
{
  TimerType = GPT_TIMER;
  TimerIndex = FspTimer::get_available_timer(TimerType);
```

```
  if (TimerIndex == 0)
  {
    FspTimer::force_use_of_pwm_reserved_timer();
    TimerIndex = FspTimer::get_available_timer(TimerType);
  }

  MyTimer.begin(TIMER_MODE_PERIODIC, TimerType, TimerIndex, freq, 0.0f,
TimerCallback);
  MyTimer.setup_overflow_irq();
  MyTimer.open();
  MyTimer.start();
}

//
// Configure segment LEDs as outputs, set timer for 200 Hz
//
void setup()
{
  for(int i = 0; i < 7; i++)
  {
    pinMode(LED[i], OUTPUT);                        // Configure as outputs
  }

  for(int i = 0; i < 4; i++)
  {
    pinMode(DIGITS[i], OUTPUT);                     // Configure as outputs
    digitalWrite(DIGITS[i], Disable);               // All digits are OFF
  }

//
// Timer1 configuration for 5ms (200Hz) interrupts
//
  StartTimer(200);                                  // Set for 200 Hz
}

//
// Main program loop. Increment count here and delay 1 second
//
void loop()
{
  cnt++;                                            // Increment count
  if(cnt > 9999) cnt = 0;                           // If 9999, reset to 0
  delay(1000);                                      // Delay 1 second
}
```

Figure 4.16: Program: **SevenSeg4**.

4.7 Project 5: 7-Segment 4-digit multiplexed LED display — reaction timer

Description: This project measures the reaction time of the user and displays it on the 7-segment display in milliseconds. The project is similar to Project 11 (Section 3.12), but here the result is sent to the 7-segment display. The on-board LED is turned ON at random times. As soon as the user sees the LED he/she is expected to press the button. The time delay between seeing the LED and pressing the button is a measure of the reaction time which is displayed by the program.

Block diagram: Figure 4.17 shows the block diagram of the project where a button, an LED, and a 4-digit multiplexed LED display are used.

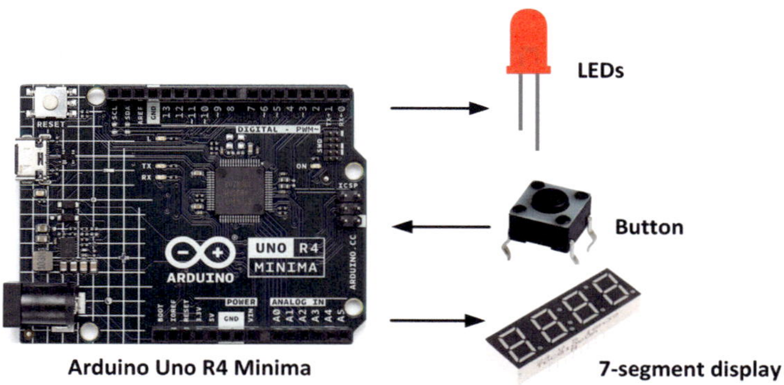

Figure 4.17: Block diagram of the project.

Circuit diagram: The circuit diagram of the project is shown in Figure 4.18. The display is connected as in the previous project. The LED and button are connected to ports 13 and 0 of the development board respectively (notice that port 13 is where the on-board LED is connected to).

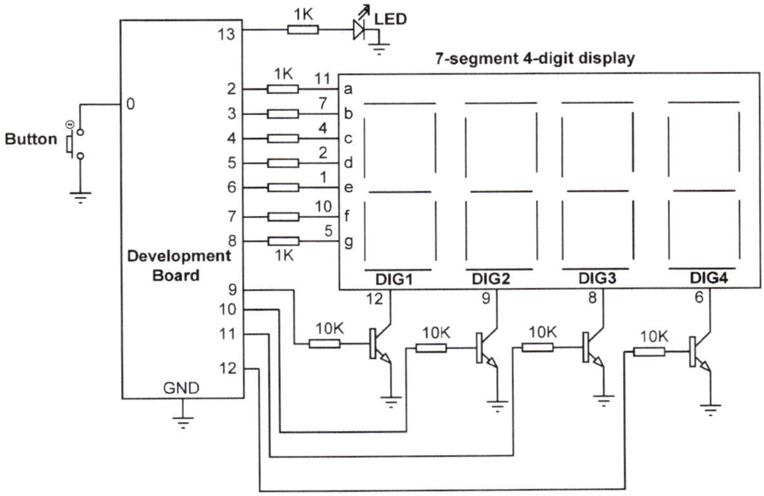

Figure 4.18: Circuit diagram of the project.

Program listing: Figure 4.19 shows the program listing (Program**: SevenReaction**). Inside the **setup()** function, the display digits, the LED, and the button are configured. The button is configured as an input with internal pullup so that normally it is at logic 1. Pressing the button changes its state to logic 0. Most parts of the program are similar to Figure 4.16. Here, inside the main program loop, a random number is generated between 1 and 20 seconds, and the program is configured to wait for some random time before turning ON the LED. At this point, the time is read by calling built-in function **millis()** and is stored in variable **StartTime**. When the button is pressed the time is read again and stored in **EndTime**. The difference between the **EndTime** and **StartTime** is the reaction time of the user. This is converted into an integer number in variable **cnt** and is displayed on the 7-segment LED. The program repeats after 5 seconds of delay.

```
//----------------------------------------------------------------------
//               7-SEGMENT 4-DIGIT DISPLAY REACTION TIMER
//               ========================================
//
// This is a 7-segment 4-digit reaction timer program. An LED is lit
// at random time. The user is expected to press the button as soon as
// he/she sees the LED turning ON. The elapsed time between the LED
// becoming ON and the user pressing the button is displayed in milliseconds.
// Maximum reaction time that can be displyed is 9999 ms
//
// Author: Dogan Ibrahim
// File  : SevenReaction
// Date  : June, 2023
//----------------------------------------------------------------------
#include "FspTimer.h"
FspTimer MyTimer;
```

```
#define ON HIGH                                           // Define ON
#define OFF LOW                                           // Define OFF
int LED[] = {8, 7, 6, 5, 4, 3, 2};                        // LEDs at ports 2 to 8
unsigned char SEG[] = {0x3F,0x06,0x5B,0x4F,0x66,          // See Table 4.1
0x6D, 0x7D, 0x07, 0x7F, 0x6F};
unsigned char DIGITS[] = {9, 10, 11, 12};                 // DIGIT ports
unsigned int MSD, m, MID2, n, MID1, LSD;
unsigned char Pattern;
int cnt = 0;
int flag = 0;
uint8_t TimerType;
int8_t TimerIndex;
int RLED = 13;
int Button = 0;
#define Enable HIGH
#define Disable LOW

//
// Group the port pins together. L is the number of bits (8 here),and No
// is the data to be displayed
//
void Display(int No, int L)
{
  int i, m, j;

  m = L - 1;
  for(i = 0; i < L; i++)
  {
    j = 1;
    for(int k = 0; k < m; k++)j = j * 2;
    if((No & j) != 0)
      digitalWrite(LED[i], ON);
    else
      digitalWrite(LED[i], OFF);
    m--;
  }
}

//
// TIMER interrupt service routine (every 5 ms, 200 Hz)
//
void TimerCallback(timer_callback_args_t __attribute((unused)) *p_args)
{
  MSD = cnt / 1000;                                      // Get MSD
  m = cnt % 1000;
```

```
  MID2 = m / 100;                                     // Get MID2
  n = m % 100;
  MID1 = n / 10;                                      // Get MID1
  LSD = n % 10;                                       // Get LSD

  if(flag == 0)
  {
    digitalWrite(DIGITS[3], Disable);                 // Disable DIG4
    Pattern = SEG[MSD];                               // Get pattern
    Display(Pattern, 7);                              // Display number
    if(cnt > 999)digitalWrite(DIGITS[0], Enable);     // Enable DIG1
    flag++;
  }
  else if(flag == 1)
  {
    digitalWrite(DIGITS[0], Disable);                 // Disable DIG1
    Pattern = SEG[MID2];                              // Get pattern
    Display(Pattern, 7);                              // Display number
    if(cnt > 99)digitalWrite(DIGITS[1], Enable);      // Enable DIG2
    flag++;
  }
  else if(flag == 2)
  {
    digitalWrite(DIGITS[1], Disable);                 // DIsable DIG2;
    Pattern = SEG[MID1];                              // Get pattern
    Display(Pattern, 7);                              // Display number
    if(cnt > 9)digitalWrite(DIGITS[2], Enable);       // Enable DIG3
    flag++;
  }
  else if(flag == 3)
  {
    digitalWrite(DIGITS[2], Disable);                 // Disable DIG3
    Pattern = SEG[LSD];                               // Get pattern
    Display(Pattern, 7);                              // Display number
    digitalWrite(DIGITS[3], Enable);                  // Enable DIG4
    flag = 0;
  }
}

//
// Get a GPT timer, set its mode, define callback ISR and start the timer
//

void StartTimer(float freq)
{
  TimerType = GPT_TIMER;
```

```
    TimerIndex = FspTimer::get_available_timer(TimerType);
    if (TimerIndex == 0)
    {
      FspTimer::force_use_of_pwm_reserved_timer();
      TimerIndex = FspTimer::get_available_timer(TimerType);
    }

    MyTimer.begin(TIMER_MODE_PERIODIC, TimerType, TimerIndex, freq, 0.0f,
  TimerCallback);
    MyTimer.setup_overflow_irq();
    MyTimer.open();
    MyTimer.start();
  }

  //
  // Set all digits OFF
  //
  void ALLOFF()
  {
   for(int i = 0; i < 4; i++)
      digitalWrite(DIGITS[i], Disable);             // All digits are OFF
  }

  //
  // Configure segment LEDs as outputs, set timer for 200 Hz
  //
  void setup()
  {
    for(int i = 0; i < 7; i++)
    {
      pinMode(LED[i], OUTPUT);                      // Configure as outputs
    }

    for(int i = 0; i < 4; i++)
    {
      pinMode(DIGITS[i], OUTPUT);                   // Configure as outputs
    }
    ALLOFF();                                       // All digits OFF

    pinMode(RLED, OUTPUT);                          // On-board LED
    digitalWrite(RLED, OFF);
    pinMode(Button, INPUT_PULLUP);                  // Button
    delay(2000);

  //
  // Timer1 configuration for 5ms (200Hz) interrupts
```

```
//
  StartTimer(200);                              // Set for 200 Hz
}

//
// Main program loop. Increment count here and delay 1 second
//
void loop()
{
  MyTimer.stop();
  ALLOFF();
  int rnd = random(1, 21);                      // Random number 1-20
  delay(rnd*1000);                              // Random delay 1-20 secs
  digitalWrite(RLED, ON);                       // LED ON
  cnt=0;                                        // Clear cnt
  float StartTime = millis();                   // Start time
  while(digitalRead(Button) == 1);              // Wait for button press
  float EndTime = millis();                     // End time
  digitalWrite(RLED, OFF);                      // LED OFF
  float ElapsedTime = EndTime - StartTime;      // Elapsed time
  cnt = int(ElapsedTime);                       // Display reaction time
  MyTimer.start();
  delay(5000);                                  // 3 seconds delay
}
```

Figure 4.19: Program: **SevenReaction**.

4.8 Project 6: Timer interrupt blinking on-board LED

Description: Now that you have learned how to use timer interrupts in your programs, in this project you will blink the on-board LED at port 13 every second using timer interrupts instead of delays. The aim of this project is to show how to generate 1-second-timer interrupts.

Program listing: In this project, you want to generate timer interrupts every second. i.e., the frequency of the interrupts is 1 Hz.

Figure 4.20 shows the program listing (Program: LEDtimer). As in the previous project timer interrupts are configured in the setup() function. The state of the LED is toggled inside the timer interrupt service routine. The main program loop does not have any code in this project.

```
//-----------------------------------------------------------------------
//          BLINKING THE ON-BOARD LED USING TIMER INTERRUPTS
//          ================================================
//
// In this project the on-board LED at port 13 is blinked every second
// using timer interrupts
//
// Author: Dogan Ibrahim
// File  : LEDtimer
// Date  : June, 2023
//-----------------------------------------------------------------------
#include "FspTimer.h"
FspTimer MyTimer;

uint8_t TimerType;
int8_t TimerIndex;
#define LED 13                                    // On-board LED at 13


//
// TIMER interrupt service routine (every sec, 1 Hz)
//
void TimerCallback(timer_callback_args_t __attribute((unused)) *p_args)
{
   digitalWrite(LED, digitalRead(LED) ^ 1);      // Toggle LED status
}

//
// Get a GPT timer, set its mode, define callback ISR and start the timer
//

void StartTimer(float freq)
{
  TimerType = GPT_TIMER;
  TimerIndex = FspTimer::get_available_timer(TimerType);
  if (TimerIndex == 0)
  {
    FspTimer::force_use_of_pwm_reserved_timer();
    TimerIndex = FspTimer::get_available_timer(TimerType);
  }

  MyTimer.begin(TIMER_MODE_PERIODIC, TimerType, TimerIndex, freq, 0.0f,
TimerCallback);
  MyTimer.setup_overflow_irq();
  MyTimer.open();
  MyTimer.start();
```

```
}

//
// Configure segment LEDs as outputs, set timer for 200 Hz
//
void setup()
{
  pinMode(LED, OUTPUT);                      // Configure as outputs
  StartTimer(1);                             // Set Timer for 1 Hz
}

void loop()
{
}
```

Figure 4.20: Program: **LEDtimer**.

Chapter 5 • Liquid Crystal Displays

5.1 Overview

In microcontroller-based systems, you usually want to interact with the system, for example, to enter a parameter, to change the value of a parameter, or to display the output of a measured variable. Data is usually entered into a system using a switch, a small keypad, or a full-blown keyboard. Data is usually displayed using an indicator such as one or more LEDs, 7-segment displays, LCDs, GLCDs, TFTs, OLEDs, etc. LCDs have the advantage that they are relatively cheap and can display alphanumeric as well as some graphical data. Some LCDs have 40 or more-character lengths, with the capability to display data in several lines. Some other LCDs can be used to display graphical images (Graphical LCDs, or simply GLCDs), such as animation. Some displays are single or multi-color, while others incorporate backlighting so that they can be viewed in dimly lit conditions.

LCDs can be connected to a microcontroller either in parallel form or through the I²C interface. Parallel LCDs (e.g. Hitachi HD44780) are connected using more than one data line and several control lines and the data is transferred in parallel form. It is common to use either 4 or 8 data lines and two or more control lines. Using a 4-wire connection saves I/O pins but it is slower since the data is transferred in two stages. I²C-based LCDs on the other hand are connected to a microcontroller using just 2 wires: the data and the clock. I²C-based LCDs are in general much easier to use and require less wiring, but they cost more than the parallel ones.

Low-level programming of LCDs is a complex task and requires a good understanding of the internal operations of the LCD controllers, including knowledge of their exact timing requirements. Fortunately, there are several libraries that can be used to simplify the use of both parallel and I²C-based LCDs.

In this chapter, you will be using the I²C-based LCD supplied with the kit in various projects. Before going into details of the LCD, it is worthwhile to first review the basic principles of the I²C bus.

5.2 The I²C bus

The I²C (or Inter-Integrated Circuit) bus was invented by Philips Semiconductor in 1982 for connecting peripheral devices and microcontrollers over short distances. The bus uses two open collector (or open-drain) bidirectional lines pulled up with resistors. **SDA** is the Serial Data line and **SCL** is the Serial Clock Line. Although the bus is bidirectional, data can only travel in one direction at any time. I²C is a bus with 7-bit address space and achieves bus speeds of 100 kbits/s in standard mode and 400 kbits/s in fast mode (faster bus speeds are also available with Version 2.0 of the bus protocol). Devices on the bus can be one or more master nodes and one or more slave nodes. The master nodes initiate the communication and generate the clock signals on the bus. Slave nodes receive the clock signals and they respond when addressed by a master.

Figure 5.1 shows an example I²C system with one master and three slaves. In a typical application, the master initiates the communication on the bus by signaling a start condition. This is followed by 7 bits of address information (10-bit addressing mode is also available), and one data direction bit, where a LOW means that the master is writing to the slave, and a HIGH means that the master is reading from the slave. With 7 bits of address, up to 128 devices can be connected to the bus. When reading and writing to the bus, you have to specify the device address, register address and the number of bytes.

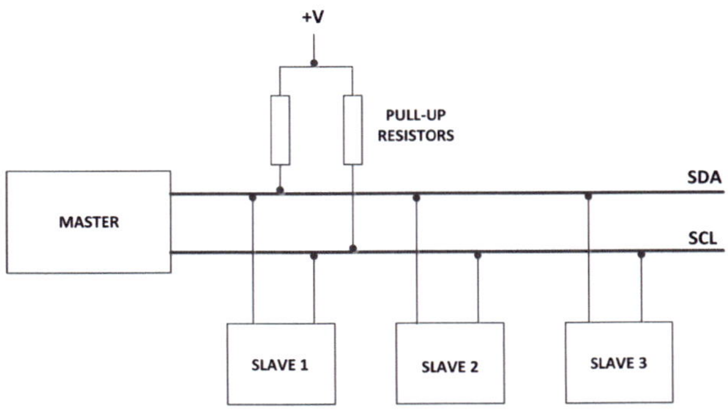

Figure 5.1: I²C bus with one master and three slaves.

5.3 I²C ports of the development board

The Arduino Uno R4 Minima development board has one I²C bus interface module, and it is available on the following pins:

A4	SDA	also marked at the end of the long header
A5	SCL	also marked at the end of the long header

The controller will send out information through the I²C bus to a 7-bit address, meaning that the technical limit of I²C devices on a single line is 128. Practically, you're never going to reach 128 devices before other limitations kick in.

Note: Pullup resistors are not mounted on the PCB but there are footprints to do so if needed.

5.4 I²C LCD

I²C-based LCD displays are normally supplied in two parts: the LCD display, and the I²C controller board. In most standard product distributions, the controller board is soldered to the back of the LCD display shown in Figure 5.2. The LCD display is basically a 1602-type parallel LCD. The controller board consists of the PCF8574 I²C controller chip (from Texas Instruments or NXP semiconductors), a small potentiometer to adjust the contrast, I/O interface pins, and address selection jumpers.

Figure 5.2: Standard I²C-based LCD.

Some distributors supply the I²C-based LCDs in two separate parts shown in Figure 5.3. The controller board is type PCF8574T from Texas Instruments, and it must be soldered to the LCD display before it can be used. The controller board has 4 pins: SCL, SDA, VCC, and GND and these must be soldered to the development board I/O pins shown in Figure 5.4.

Figure 5.3: I²C LCD parts included in the kit.

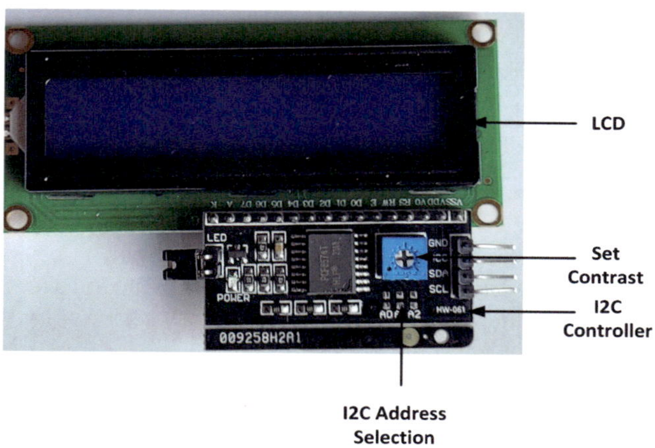

Figure 5.4: Solder the I²C controller board to the LCD.

As shown in Figure 5.5, on some LCDs the I²C address of the PCF8574T chip is selected by 3 jumpers labelled A0, A1 and A2 on the controller board. By default, the address is set to 0x27 (i.e., no jumper connections).

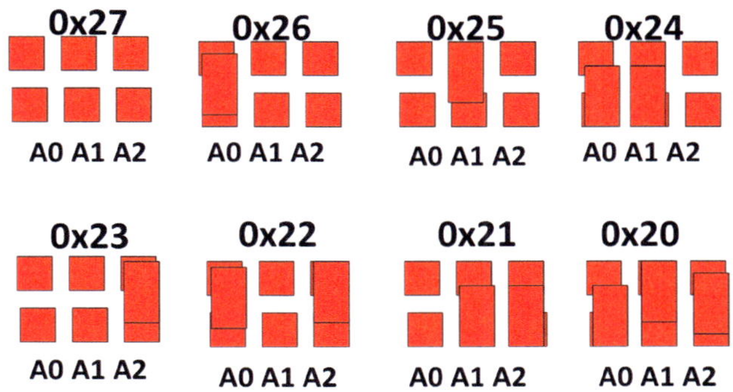

Figure 5.5: I²C address selection.

Before using the I²C LCD, you have to add the I²C library to your IDE. Libraries are often distributed as a ZIP file or folder where the name of the folder is the name of the library. Inside the folder, there is a **.cpp** file, a **.h** file, a **keywords.txt** file, **examples** folder, and other files that may be required by the library. You should not unzip the library.

The steps to add the I²C LCD library are as follows:

- Click to open the **LIBRARY MANAGER**.
- Type **i2c lcd** in the search box.
- Scroll down to find the library: **LCD_I2C by Blackhack**.
- Click **INSTALL** to install the library.
- Close the **LIBRARY MANAGER**.

At the time of writing this book the library version was 2.3.0.

To test that the library has been added successfully, enter the following lines in a newly created program:

```
#include <LCD_I2C.h>
LCD_I2C lcd(0x27, 16, 2);

void setup()
{
   lcd.begin();
}

void loop()
{
}
```

- Compile the program. There should be no errors.

The I²C LCD library supports many functions. Some most commonly used functions are:

begin():	**initialize LCD (This must be the first function call)**
clear():	clear the screen
home():	home the cursor
noBlink():	stop blinking cursor
blink():	enable blinking cursor
noCursor():	hide cursor
cursor():	display cursor
scrollDisplayLeft():	scroll display left
scrollDisplayRight():	scroll display right
noBacklight():	disable backlight
backlight():	enable backlight
setCursor(column, row):	set cursor position (0, 0) is the top left position
print():	print data on LCD

The address of the I²C LCD must be defined at the beginning of the program. For example, if the address is 0x27 and the LCD is 16 columns by 2 rows (i.e., 16 × 2), then:

```
LCD_I2C lcd(0x27, 16, 2);
```

Following the above statement, you can call the LCD functions by indexing them with the keyword **lcd**. For example, to initialize the LCD:

```
lcd.begin();                    // initialize the lcd
```

or, for example to enable the backlight, use:

```
lcd.backlight();                // Enable backlight
```

5.5 Project 1: Display text on the LCD

Description: In this project, you will display the text **MY LCD** at the top row (row 0), starting from column position 5 of the LCD. The aim of this project is to show how the I²C LCD can be connected and used in a program. The project also shows how to display text on the LCD.

Block diagram: Figure 5.6 shows the block diagram of the project.

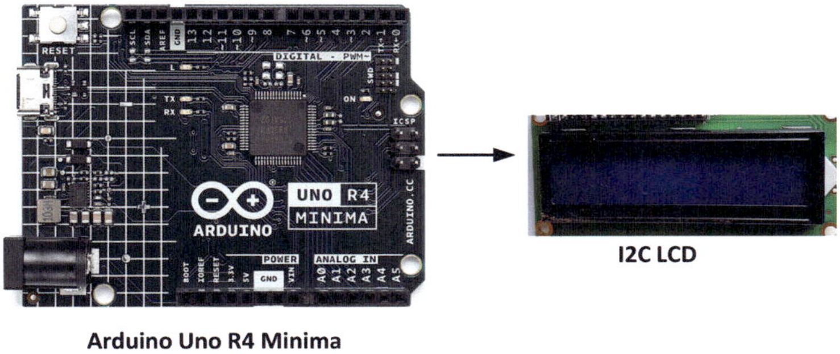

Figure 5.6: Block diagram of the project.

Circuit diagram: The I²C LCD is connected to the SDA and SCL pins of the Arduino Uno R4 Minima development board shown in Figure 5.7.

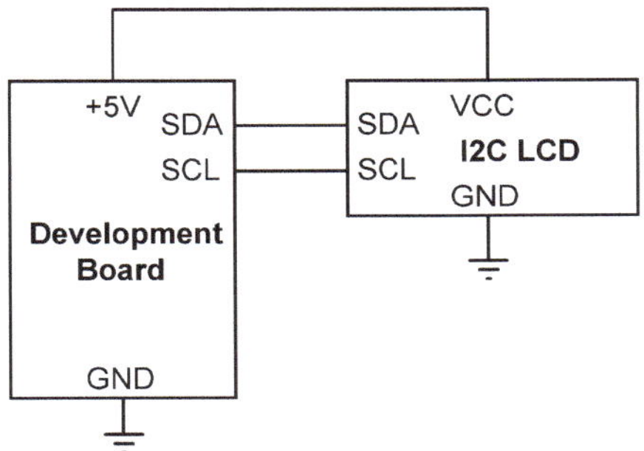

Figure 5.7: Circuit diagram of the project.

Program listing: Figure 5.8 shows the program listing (Program: **LCD1**). The LCD library is initialized inside the **setup()** function and the backlight is turned ON. Inside the main program loop, the text **MY LCD** is displayed at row 0, column 5 of the LCD. Notice that (0, 0) is the top-left corner (i.e. home) position of the cursor. Rows are numbered 0 and 1, and columns 0 to 15.

```
//----------------------------------------------------------------------
//          DISPLAYING TEXT ON THE LCD
//          ==========================
//
// In this project the text MY LCD is displayed at the top row,
// starting column 5 of the LCD
//
// Author: Dogan Ibrahim
// File  : LCD1
// Date  : June, 2023
//----------------------------------------------------------------------
#include <LCD_I2C.h>
LCD_I2C lcd(0x27, 16, 2);
```

```
void setup()
{
  lcd.begin();                                    // initialize the lcd
  lcd.backlight();                                // Backlight ON
}


void loop()
{
  lcd.setCursor(5, 0);                            // Row 0, Column 5
  lcd.print("MY LCD");                            // Display text

  while(1);                                       // Wait here forever
}
```

Figure 5.8: Program: **LCD1**.

Construction: Figure 5.9 shows the project where the I²C LCD pins are connected to the development board using the supplied jumper wires.

Figure 5.9: Construction of the project.

5.6 Project 2: Scrolling text on the LCD

Description: In this project, you will initially display text **SCROLL** at the top row (row 0), starting from column position 10 of the LCD. This text is then scrolled left by 10 cursor positions with 500 ms between each scroll. After a delay of 2 seconds, the same text is scrolled right this time again by 10 positions. This process is repeated until stopped by the user.

The aim of this project is to show how text can be scrolled on the display.

The block diagram and circuit diagram of the project are in Figure 5.6 and Figure 5.7 respectively.

Program listing: Figure 5.10 shows the program listing (Program: **LCDScroll**). LCD library functions **scrollDisplayLeft()** and **scrollDisplayRight()** are used in the program.

```
//-----------------------------------------------------------------------
//          SCROLL TEXT ON THE LCD
//          ======================
//
// In this project the text SCROLL is scrolled left and right
//
// Author: Dogan Ibrahim
// File  : LCDScroll
// Date  : June, 2023
//-----------------------------------------------------------------------
#include <LCD_I2C.h>
LCD_I2C lcd(0x27, 16, 2);

void setup()
{
  lcd.begin();                               // initialize the lcd
  lcd.backlight();                           // Backlight ON
}


void loop()
{
  lcd.setCursor(10, 0);                      // Row 0, Column 10
  lcd.print("SCROLL");                       // Display text

  for(int i=0; i < 10; i++)                  // Scroll LEFT
  {
    lcd.scrollDisplayLeft();
    delay(500);
  }

  delay(2000);                               // Delay 2 seconds
  for(int i = 0; i < 10; i++)                // Scroll RIGHT
  {
    lcd.scrollDisplayRight();
    delay(500);
  }

  delay(2000);                               // Delay 2 seconds
}
```

Figure 5.10: Program: **LCDScroll***.*

5.7 Project 3: Display custom characters on the LCD

Description: The LCD display has two types of memory, called CGROM and CGRAM. CGROM memory is non-volatile and stores all permanent fonts used by the LCD and this memory cannot be modified. CGRAM memory on the other hand is volatile and can be modified by users to store user-defined characters.

In this project, you will generate an up-arrow character and then display it on your LCD.

Generating custom characters

Generating a custom character requires drawing the required character on a 5 × 8-pixel grid and then calculating the corresponding bit pattern. There are several tools on the Internet that can be used to create your own characters. In this project, the Custom Character Generator at the following website is used to generate the up arrow:

https://lastminuteengineers.com/i2c-lcd-arduino-tutorial/

To generate your own character simply click on the grid. Figure 5.11 shows the character generated by the author and the corresponding bit pattern.

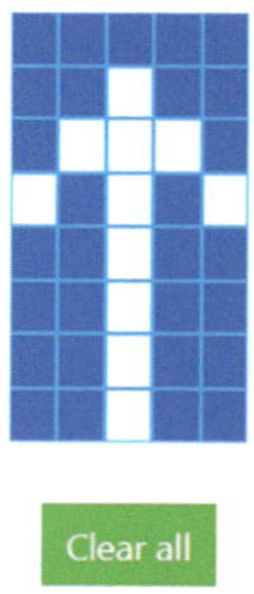


```
byte Character[8] =
{
0b00000,
0b00100,
0b01110,
0b10101,
0b00100,
0b00100,
0b00100,
0b00100
};
```

Figure 5.11: Generated arrow character and its bit pattern.

You can now copy the generated code into your program and display the required character.

Program listing: Figure 5.12 shows the program listing (Program: **LCDArrow**). The generated array name is changed to **UpArrow**. Inside the **setup()** function, the character is loaded into LCD memory with index 0. Inside the main program loop, the cursor is set to row 0, column 5, and the loaded character is displayed by calling the LCD function **write()** with index 0. Figure 5.13 shows the displayed character.

```
//----------------------------------------------------------------------
//          USING A CUSTOM CHARACTER
//          ========================
//
// In this project an up arrow character is generated and displad on LCD
//
// Author: Dogan Ibrahim
```

```
// File   : LCDArrow
// Date   : June, 2023
//------------------------------------------------------------------
#include <LCD_I2C.h>
LCD_I2C lcd(0x27, 16, 2);

byte UpArrow[8] =
{
0b00000,
0b00100,
0b01110,
0b10101,
0b00100,
0b00100,
0b00100,
0b00100
};


void setup()
{
  lcd.begin();                              // initialize the lcd
  lcd.backlight();                          // Backlight ON
  lcd.createChar(0, UpArrow);               // Create character with index 0
  lcd.clear();                              // Clear display
}


void loop()
{
  lcd.setCursor(5, 0);                      // Row 0, Column 5
  lcd.write(0);                             // Display the character

  while(1);                                 // Wait here forever
}
```

Figure 5.12: Program: **LCDArrow***.*

Figure 5.13: Generated character displayed on the LCD.

5.8 Project 4: LCD based conveyor belt goods counter

Description: In this project, you count the number of items (e.g., bottles) passing on a conveyor belt and display the result continuously on the LCD.

Block diagram: Figure 5.14 shows the system. It is assumed that you wish to count the number of bottles passing on a conveyor belt. A light beam is directed at a point to the passing bottles. At the other side of the beam, a light-dependent resistor (LDR) is used to detect when the light beam is interrupted. When this happens, a signal is sent to the development board which then increments a counter and displays the total count on the LCD.

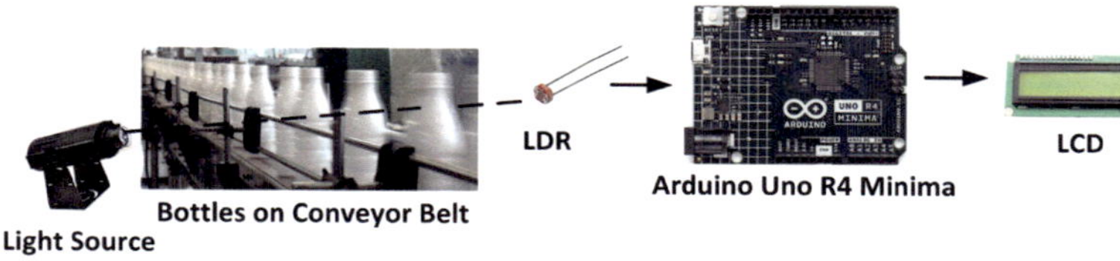

Figure 5.14: Block diagram of the project.

Light-dependent resistor

A light-dependent resistor is simply a resistor whose resistance changes with the application of light to its surface (Figure 5.15). Although you will be using only one in this project, there are 3 LDRs supplied with the kit. The resistance of an LDR increases as the light intensity falling on the device is reduced (see Figure 5.15).

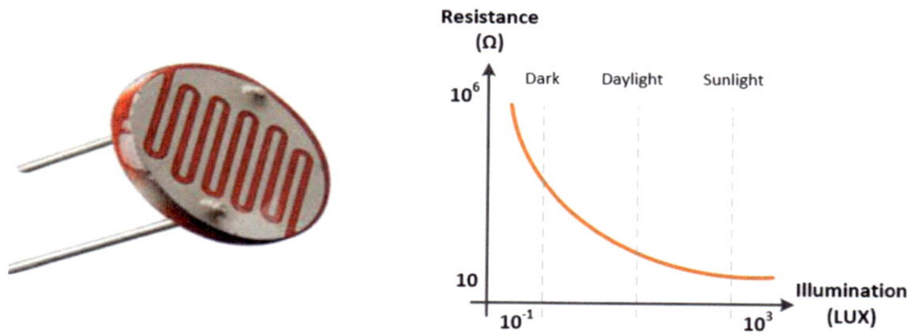

Figure 5.15: LDR and its typical characteristics.

LDRs are usually used in series with resistors in a circuit to form a resistive potential divider circuit. The voltage at the output of the potential divider circuit is used to send a trigger signal when the light intensity is below (or above) a set level. The trigger point can be detected by an MCU using both analog and digital inputs.

Circuit diagram: Figure 5.16 shows the circuit diagram of the project. The supplied potentiometer (about 5 kΩ) is used in series with the LDR. The potentiometer is adjusted such that the output voltage goes over 3 V when the light falling on the LDR is reduced by a passing bottle. The output of the potential divider circuit is fed to digital port 2 of the development board.

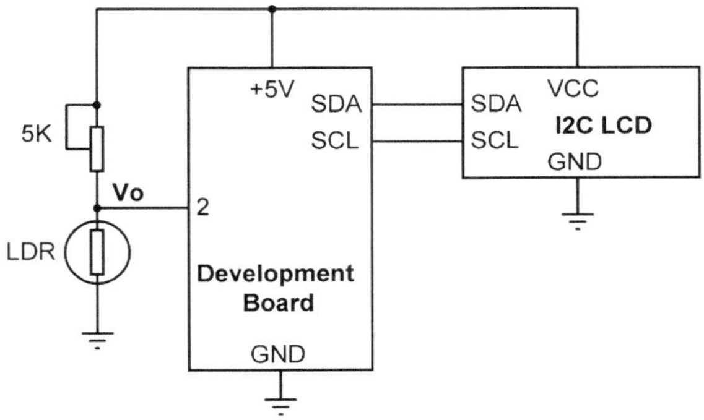

Figure 5.16: Circuit diagram of the project.

It was measured by the author that the resistance of the supplied LDR is about 0.5 kΩ in light and increases to about 100 kΩ when in the dark. Assuming the potentiometer is set to its middle arm point, i.e., 2.5 kΩ:

The voltage at the output when light falls on the LDR is:

V_o = 5 V × 0.5 kΩ / (2.5 kΩ + 0.5 kΩ) = 0.8 V which is at logic 0.

Similarly, the voltage at the output when it is dark is:

V_o = 5 V × 100 kΩ / (2.5 kΩ + 100 kΩ) = 4.87 V which is at logic 1.

You will set the potentiometer arm to just below its mid-point so that the output voltage is even lower than 0.8 V when it is light. The exact point can easily be determined by experimentation.

Program listing: Figure 5.17 shows the program listing (Program: **LDRConveyor**). At the beginning of the program, the LCD library is defined and LDR is assigned to port 2. Inside the **setup()** function, LDR is configured as an input, LCD is initialized and the heading **TOTAL COUNT** is sent to the top row of LCD. Inside the main program loop, the cursor is set to (0, 1), and the program waits while there is light on the LDR (i.e., no bottle detected). When a bottle is detected, the program comes out of the **while** statement and increments the total count (variable **Total**). This variable is converted into a string (just in case it is required to display text as well) and is displayed on the LCD. The program then waits until the bottle passes in front of the LDR (i.e., while the output voltage Vo is 1). This cycle is repeated after a small delay.

```
//------------------------------------------------------------------------
//          CONVEYOR BELT GOODS COUNTER
//          ===========================
//
// In this project an LDR is used togther with a potentiometer to count
// the number of goods (e.g. bottles) passing on a conveyor belt. The
// total count is displayed on the LCD
//
// Author: Dogan Ibrahim
// File  : LDRConveyor
// Date  : June, 2023
//------------------------------------------------------------------------
#include <LCD_I2C.h>
LCD_I2C lcd(0x27, 16, 2);

unsigned int Total = 0;                     // Total count
int LDR = 2;                                // LDR at pin 2

void setup()
{
  pinMode(LDR, INPUT);                      // LDR is input
  lcd.begin();                              // initialize the lcd
  lcd.backlight();                          // Backlight ON
  lcd.clear();                              // Clear LCD
  lcd.setCursor(0, 0);
  lcd.print("  TOTAL COUNT");               // Display heading
  delay(100);                               // Wait a bit
}


void loop()
{
  lcd.setCursor(0, 1);                      // Row 1, Column 0
  while(digitalRead(LDR) == 0);             // Wait if light (no bottle)
  Total++;                                  // Increment Total (bottle detected)
  String Tot = String(Total);               // Convert Total to string
  lcd.print(Tot);                           // Display total
  while(digitalRead(LDR) == 1);             // Wait until bottle passes
  delay(100);                               // A bit of delay
}
```

Figure 5.17: Program: **LDRConveyor**.

5.9 Project 5: LCD based accurate clock using timer interrupts

Description: This is an accurate clock project with an LCD display. The clock has 3 buttons to set the time as follows:

SET: press to set the time or return back to clock mode.
HRS: press to set the hours. Hours is incremented by one every time it is pressed.
MINS: press to set the minutes. Minutes is incremented by one every time it is pressed.

The clock is interrupt-based where a timer is used to generate interrupts every second. The time is then set according to these interrupts. Button **SET** is external interrupt based, so pressing this button at any time puts the clock into set mode where the hours and minutes can be set. Seconds are set to 0 on exit from the SET mode. Pressing the button while in this mode puts the clock back into operational mode.

Block diagram: Figure 5.18 shows the block diagram of the project.

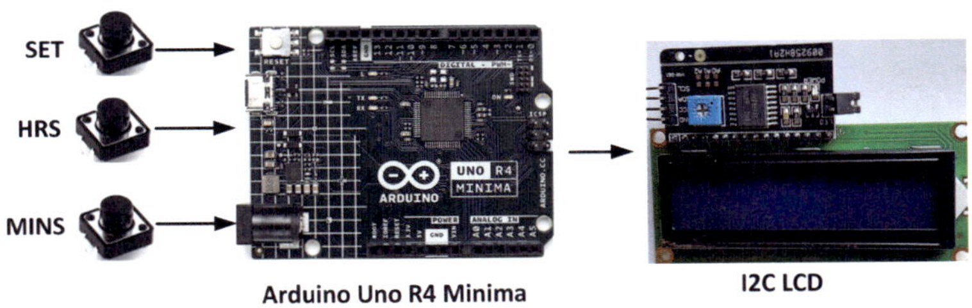

Figure 5.18: Block diagram of the project.

Circuit diagram: The circuit diagram is shown in Figure 5.19. Buttons **SET**, **HRS**, and **MINS** are connected to port pins 2, 3 and 4, respectively (remember that external interrupt is available on port pin 2) and are pulled up in the software. The LCD is connected as in the previous LCD projects via the SDL and SDA pins.

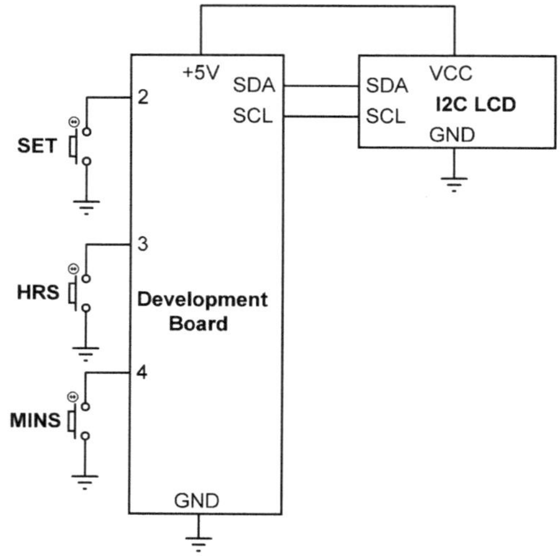

Figure 5.19: Circuit diagram of the project.

Program listing: Figure 5.20 shows the program listing (Program: **CLOCK**). At the beginning of the program, buttons **SET**, **HRS** and **MINS** are assigned to port numbers 2, 3 and 4 respectively. Inside the **setup()** function, the buttons are configured as inputs and are pulled up in the software by internal resistors. The LCD is initialized and the backlight is turned ON. Then button **SET** is configured as an external interrupt pin on the falling edge by calling function **attachInterrupt**. The external interrupt service routine **SETbutton()** is called whenever button **SET** is pressed. Also, inside the **setup()** function, timer interrupts are configured to generate interrupts at every second.

The timer interrupt service routine is the function named **ISR_TIMER1_COMPA_vect()**. Inside this routine, seconds are incremented by one. Then the minutes and hours are updated accordingly. The code inside the timer-interrupt service routine runs only if the clock is in operational mode (i.e. the **SET** button is not pressed).

Inside the main program loop, the time is displayed on the LCD in the format HH:MM:SS if the clock is in operational mode.

Pressing the **SET** button puts the clock into set mode. In this mode **HRS:** and **MINS:** are displayed on the second row of the LCD. Pressing the **HRS** button increments hours by one. Similarly, pressing the **MINS** button increments minutes by one. If either of these buttons are kept pressed, then their values are incremented continuously. When you are happy with the hours and minutes settings, you should press the **SET** button to return to the clock mode. The clock continues to run from the set hours and minutes with the seconds set to 0.

```
//--------------------------------------------------------------------
//          ACCURATE CLOCK WITH LCD DISPLAY
//          ===============================
//
// This is a clock project with LCD display. 1 second Timer interrupts
// are used for timing. Three buttons are used to set the clock as
// described in the text of the project
//
// Author: Dogan Ibrahim
// File  : CLOCK
// Date  : June, 2023
//--------------------------------------------------------------------
#include <LCD_I2C.h>
#include "FspTimer.h"

FspTimer MyTimer;
LCD_I2C lcd(0x27, 16, 2);

uint8_t TimerType;
int8_t TimerIndex;

int SET = 2;                                    // SET button at port 2
int HRS = 3;                                    // HRS button at port 3
int MINS = 4;                                   // MINS button at port 4

volatile int hours = 0, minutes = 0, seconds = 0;
bool SetTime = false;                           // Operating mode

void setup()
{
  pinMode(SET, INPUT_PULLUP);                   // SET is input
  pinMode(HRS, INPUT_PULLUP);                   // HRS is input
  pinMode(MINS, INPUT_PULLUP);                  // MINS is input
  lcd.begin();                                  // initialize the lcd
  lcd.backlight();                              // Backlight ON
  lcd.clear();                                  // Clear LCD

//
// Configure SET button for external interrupts
//
  attachInterrupt(digitalPinToInterrupt(SET), SETbutton, FALLING);

//
// Configure Timer 1 for 1 second interrupts (1 Hz)
//
  StartTimer(1);
```

```
}

//
// TIMER interrupt service routine (every second)
// Time is only updated if we are in operational mode
//
void ISR(timer_callback_args_t __attribute((unused)) *p_args)
{
  if(!SetTime)                                    // If operational
  {
    seconds++;                                    // Increment seconds
    if(seconds == 60)
    {
      seconds = 0;
      minutes++;                                  // Increment minutes
      if(minutes == 60)
      {
        minutes = 0;
        hours++;                                  // Increment hours
        if(hours == 24)
        {
          hours = 0;
        }
      }
    }
  }
}

//
// Get a GPT timer, set its mode, define callback ISR and start the timer
//
void StartTimer(float freq)
{
  TimerType = GPT_TIMER;
  TimerIndex = FspTimer::get_available_timer(TimerType);
  if (TimerIndex == 0)
  {
    FspTimer::force_use_of_pwm_reserved_timer();
    TimerIndex = FspTimer::get_available_timer(TimerType);
  }

  MyTimer.begin(TIMER_MODE_PERIODIC, TimerType, TimerIndex, freq, 0.0f, ISR);
  MyTimer.setup_overflow_irq();
  MyTimer.open();
  MyTimer.start();
}
```

```
//
// External interrupt service routine via the SET pin
//
void SETbutton()
{
  SetTime = !SetTime;                        // Toggle SetTime

}

//
// Main program loop
//
void loop()
{
  if(!SetTime)                               // If in operational mode
  {
    lcd.setCursor(0, 1);                     // At second row
    lcd.print("                ");           // Clear second row of LCD
    lcd.setCursor(0, 0);                     // To row 0, column 0
    if(hours < 10)lcd.print("0");            // Display leading 0
    lcd.print(hours);                        // Display hours
    lcd.print(":");                          // Display :

    if(minutes < 10)lcd.print("0");          // Display leading 0
    lcd.print(minutes);                      // Display minutes
    lcd.print(":");                          // Display :

    if(seconds < 10)lcd.print("0");          // Display leading 0
    lcd.print(seconds);                      // Display seconds
    lcd.print(" ");
  }
  else                                       // In SET mode
  {
    lcd.setCursor(0, 1);
    lcd.print("                ");
    lcd.setCursor(0, 1);

    lcd.print("HRS:");                       // Display HRS:
    lcd.print(hours);
    if(digitalRead(HRS) == 0)hours++;        // Increment hours
    if(hours > 23)hours = 0;

    lcd.setCursor(9, 1);
    lcd.print("MINS:");                      // Display MINS:
    lcd.print(minutes);
```

```
    if(digitalRead(MINS) == 0)minutes++;          // Increment minutes
    if(minutes > 59)minutes = 0;
    seconds = 0;                                   // Set seconds to 0
    delay(250);
  }
}
```

Figure 5.20: Program: ***Clock****.*

Figure 5.21a shows the clock running in operational mode. When the **SET** button is pressed the clock enters the set mode shown in Figure 5.21b. In this mode, pressing the **HRS** and **MINS** buttons set the clock as required. Pressing the **SET** button again puts the clock back into operational mode as in Figure 5.21a.

Figure 5.21: Operational mode and set mode.

5.10 Project 6: LCD dice

Description: This is an LCD-based dice project. When a button is pressed two random numbers between 1 and 6 are displayed on the LCD. The numbers are displayed for 3 seconds and after this time the display is cleared, ready to generate new numbers. Message **READY...** is displayed when the program is ready.

Block diagram: Figure 5.22 shows the block diagram of the project.

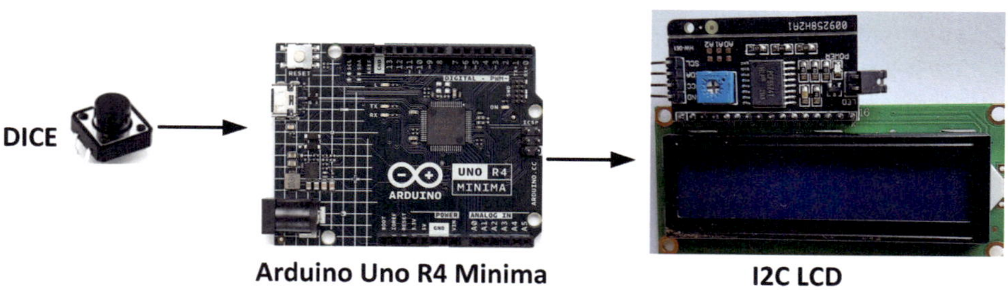

Figure 5.22: Block diagram of the project.

Circuit diagram: Figure 5.23 shows the circuit diagram of the project. The button is connected to port 2.

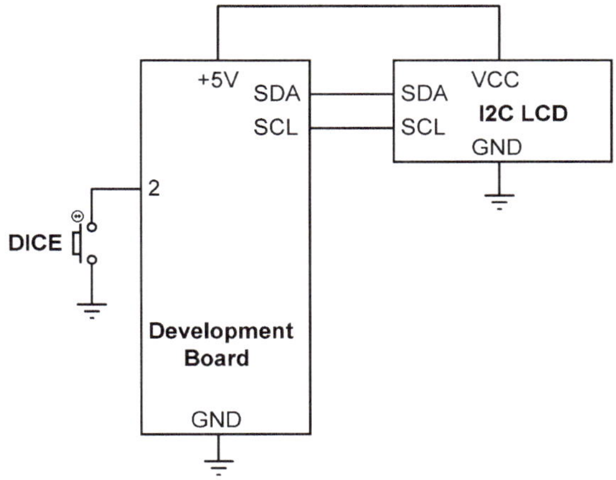

Figure 5.23: Circuit diagram of the project.

Program listing: The program listing is shown in Figure 5.24 (Program:**LCDDICE**). Button DICE is assigned to port 2 and is configured as an input. Random number generator seed is loaded from analog input A0. Since this input is floating it is expected that every time you call it may have a different value so that a different set of random numbers will be generated. Two random numbers are generated between 1 and 6 and are displayed on the LCD.

```
//--------------------------------------------------------------------------
//                          LCD DICE
//                          ========
//
// This is an LCD based dice. Two numbers are generated and displayed
// on LCD when the button is pressed
//
// Author: Dogan Ibrahim
// File  : LCDDICE
// Date  : June, 2023
//--------------------------------------------------------------------------
#include <LCD_I2C.h>
LCD_I2C lcd(0x27, 16, 2);

int DICE = 2;                                        // DICE button at port 2

void setup()
{
  pinMode(DICE, INPUT_PULLUP);                       // DICE is input
  lcd.begin();                                       // initialize the lcd
  lcd.backlight();                                   // Backlight ON
  lcd.clear();                                       // Clear LCD
```

```
  randomSeed(analogRead(0));                        // Random seed
}

void loop()
{
 lcd.setCursor(0, 0);
 lcd.print("READY...");
 while(digitalRead(DICE) == 1);                     // Wait until pressed
 int r1 = random(1, 7);                             // Generate 1 - 6
 int r2 = random(1, 7);                             // Another number
 lcd.clear();                                       // Clear LCD
 lcd.setCursor(0, 0);
 lcd.print(r1);                                     // Display first number
 lcd.print("     ");
 lcd.print(r2);                                     // Display second number
 delay(3000);
 lcd.clear();
}
```

Figure 5.24: Program: **LCDDICE**.

Chapter 6 • Sensors

6.1 Overview

In the last chapter, you explored the use of LC displays (LCDs) in projects. There are many types of sensors supplied with the kit. In this chapter, you will learn how to use some of these sensors in various interesting projects. Some other sensors and actuators will be covered in later chapters of the book.

6.2 Project 1: Analog temperature sensor

Description: In this project, you will be using the LM35 analog temperature sensor chip supplied with the kit. The project will measure the ambient temperature and then display it on LCD, updating every 5 seconds.

Block diagram: Figure 6.1 shows the block diagram of the project which consists of the LM35 sensor chip, the development board, and the I²C LCD.

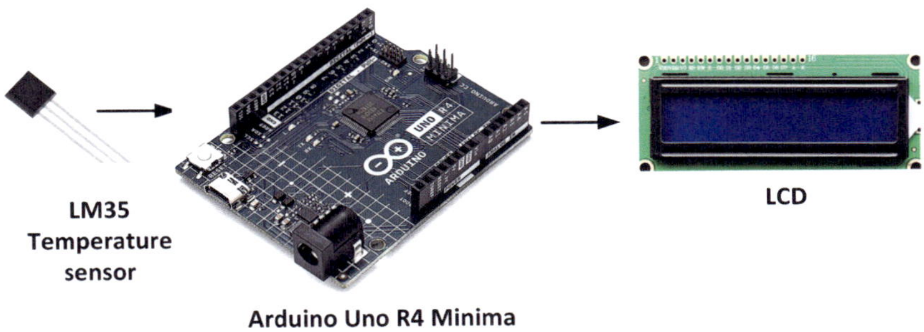

Figure 6.1: Block diagram of the project.

Circuit diagram: LM35 is a 3-pin temperature sensor chip (Figure 6.2) with pins: +V, GND, and OUT. The OUT pin is connected to analog input **A0** of the Arduino Uno R4 development board. The analog-to-digital converter (ADC) on the processor is by default 10-bits wide, having 1024 quantization levels. Therefore, with a +5 V reference voltage, each quantization level corresponds to 4.88 mV. For example, 4.88 mV analog input corresponds to 10-bit digital data: 00 0000 0001, similarly, 9.76 mV corresponds to 00 0000 0010, and so on. The reading of the ADC must be multiplied by 1023/5000 to give the actual physical voltage in millivolts present at the analog input.

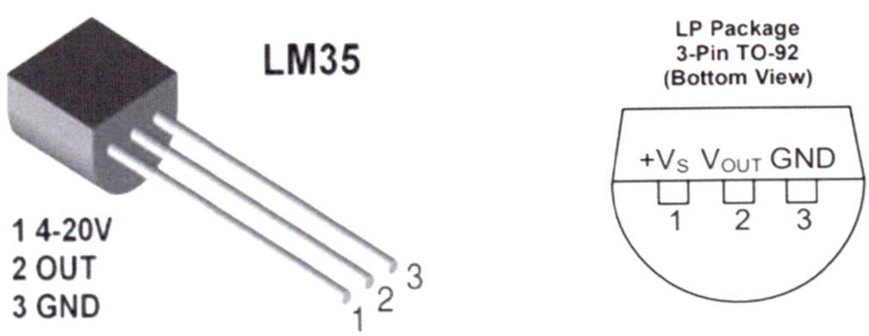

Figure 6.2: LM35 temperature sensor chip.

The LM35 temperature sensor has the following basic specifications:

- Calibrated in degrees celsius
- Linear 10 mV/ºC output
- 0.5 ºC accuracy
- −55 ºC to +125 ºC operation
- Operation from +4 V to +30 V
- Less than 60 µA current drain
- Low self-heating (0.08 ºC in still air)

The output voltage of LM35 is linearly proportional to the measured temperature and is given by:

$$T = V_o / 10$$

Where T is the measured temperature in degrees C, and Vo is the sensor output voltage in millivolts. For example, 250 mV output corresponds to 25 ºC; 300 mV corresponds to 30 ºC, and so on. For simplicity, to find the measured temperature, divide the analog voltage read in millivolts by 10.

Figure 6.3 shows the circuit diagram of the project.

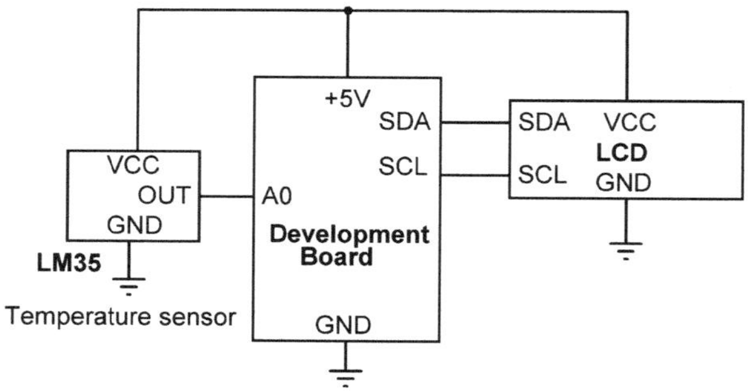

Figure 6.3: Circuit diagram of the project.

Program listing: The program listing is shown in Figure 6.4 (Program: **LM35**). At the beginning of the program, the I²C LCD header file is included and LM35 is assigned to analog input port A0 of the development board. The temperature is read inside the main program loop, converted into degrees C and displayed on the LCD every 2 seconds. *Notice that you might get fluctuation values. It is recommended by the manufacturers to use an RC filter circuit at the output of the LM35 for stable operation (see manufacturers data sheet).*

```
//----------------------------------------------------------------------
//                   LM35 TEMPERATURE DISPLAY
//                   ========================
//
// In this project the LM35 temperature sensor chip is used to measure
// and display the ambient temperature on the LCD
//
// Author: Dogan Ibrahim
// File  : LM35
// Date  : June, 2023
//----------------------------------------------------------------------
#include <LCD_I2C.h>
LCD_I2C lcd(0x27, 16, 2);

#define LM35 A0                                 // LM35 at port A0
int raw;

void setup()
{
    lcd.begin();                                // Initialize LCD
    lcd.backlight();                            // BAcklight ON
}

void loop()
{
  raw = analogRead(LM35);                       // Read temperature
  float mV = 5000.0 * raw / 1023.0;             // in mV
  float Temperature = mV / 10.0;                // True temperature
  lcd.setCursor(0, 0);
  lcd.clear();                                  // Clear LCD
  lcd.print("T = ");                            // Display T =
  lcd.print(Temperature);                       // Display temperature
  lcd.print(" C");                              // Display C
  delay(2000);                                  // Wait 2 seconds
}
```

Figure 6.4: Program: **LM35.**

Note: The default ADC resolution of the Arduino Uno R4 processor is 10 bits. The processor supports resolutions up to 14 bits for much more accuracy. You can change the resolution to 12 bits or to 14 bits. To change to 14 bits, enter the following statement in your **setup()** function:

```
analogReadResolution(14);
```

6.3 Project 2: Voltmeter

Description: This is a voltmeter project. The voltage to be measured is applied to an analog input and its value in millivolts is displayed on the LCD. The maximum allowable input voltage is +5 V. It is shown later in the project how to use a resistive potential divider circuit to extend the range of the voltmeter.

Circuit diagram: The voltage to be measured is applied to analog input A0 shown in Figure 6.5. The I²C LCD is connected as in the previous project.

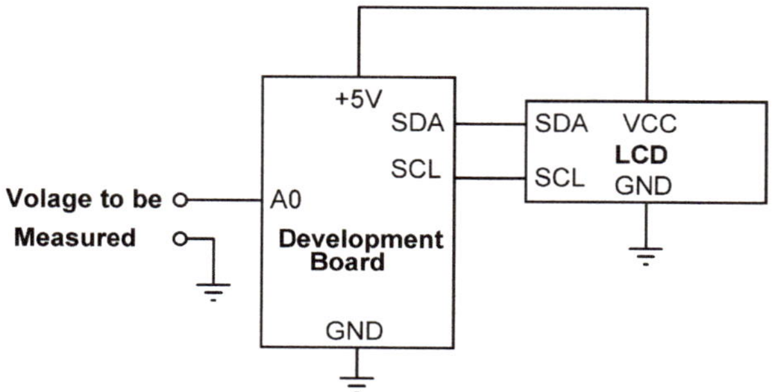

Figure 6.5: Circuit diagram of the project.

Program listing: Figure 6.6 shows the program listing (Program: **Voltmeter**). The program is very simple. The input voltage is read and displayed in millivolts on the LCD.

```
//-----------------------------------------------------------------------
//                        VOLTMETER
//                        =========
//
// This is a voltmeter project which displays te input voltage in mV
//
// Author: Dogan Ibrahim
// File  : Voltmeter
// Date  : June, 2023
//-----------------------------------------------------------------------
#include <LCD_I2C.h>
LCD_I2C lcd(0x27, 16, 2);

#define voltmeter A0                                 // Input at port A0
int raw;

void setup()
{
    lcd.begin();                                     // Initialize LCD
    lcd.backlight();                                 // Backlight ON
}
```

```
void loop()
{
  raw = analogRead(voltmeter);                    // Read temperature
  float mV = 5000.0 * raw / 1023.0;               // in mV
  lcd.setCursor(0, 0);
  lcd.clear();                                    // Clear LCD
  lcd.print("V = ");                              // Display V =
  lcd.print(mV);                                  // Display voltage
  lcd.print(" mV");                               // Display C
  delay(1000);                                    // Wait 1 second
}
```

Figure 6.6: Program: ***Voltmeter****.*

Extending the voltmeter range

The useful range of your voltmeter can easily be extended by using resistive potential divider circuits. For example, the circuit shown in Figure 6.7 attenuates the input voltage by a factor of 4 so that input voltages up to 20 V can be measured. You should of course have to multiply the readings by 4.

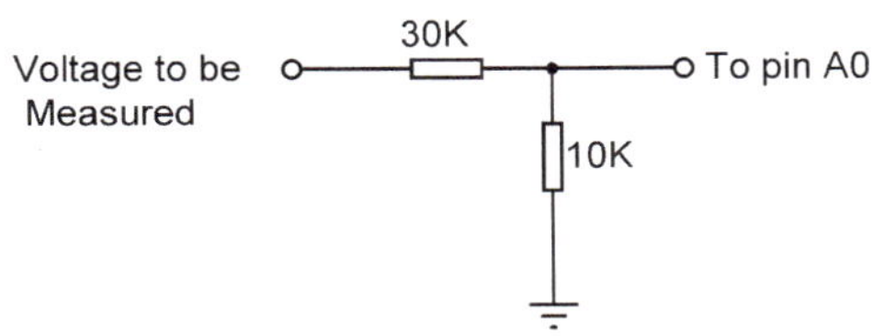

Figure 6.7: Resistive potential divider circuit.

6.4 Project 3: On/off temperature controller

Description: Temperature control is very important in many industrial, commercial, and domestic applications. The success of many chemical reactions depends on applying the correct temperature. Most temperature control systems are feedback based, where the temperature of the place whose temperature is to be controlled is measured and compared with the desired temperature. Then, an algorithm is used to achieve the desired temperature. Most professional temperature control systems are based on PID (Proportional+Integral+Derivative) type of feedback control algorithms. Another simplified type of control is the ON-OFF type of control. Here, the temperature is measured (**RoomTemp**) and compared with the desired temperature (**SetTemp**). If the measured temperature is lower than the desired one, then a heater is turned ON to increase the temperature. If on the other hand, the measured temperature is higher than the desired temperature, then the heater is turned OFF and additionally, a fan can be used to assist to lower the temperature. In ON-OFF type of control applications, relays are usually used to activate/deactivate the heater supply voltage. The main disadvantage of the ON-OFF type of temperature control is that it is not possible to achieve very accurate temperature control. Also, the relay can wear out as it has to operate many times, unless a semiconductor type of relay is used.

In this project, you will be implementing an ON-OFF type of control algorithm to control the temperature of a room. In this project, the **SetTEmp** is fixed in the program for simplicity. A red LED is connected to the development board to indicate when the relay (i.e. the heater) is ON.

Block diagram: Figure 6.8 shows the block diagram of the project. Note that the LM35, Relay, and I²C LCD are all included in the kit. The heater is not included in the kit.

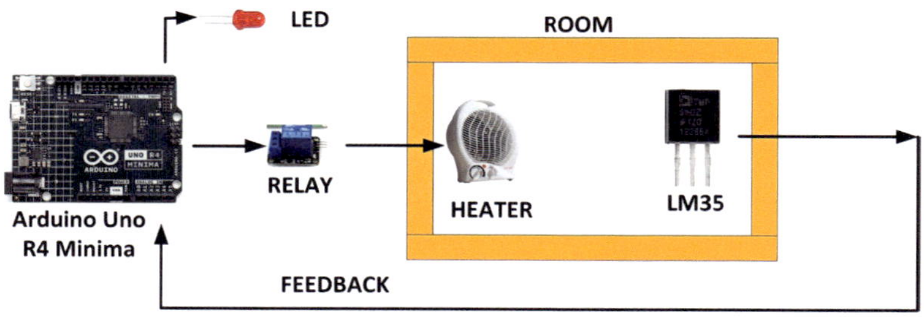

Figure 6.8: Block diagram of the project.

Circuit diagram: The circuit diagram of the project is shown in Figure 6.9. The interface between the development board and the external components is as follows (ground and voltage supply pins are not shown):

External component	**Development board port**
LM35	A0
RELAY (Pin S)	2
LED	3
LCD	SDA, SCL

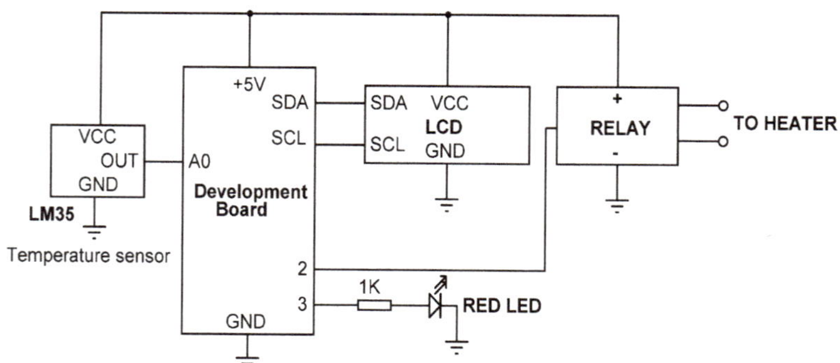

Figure 6.9: Circuit diagram of the project.

Note that some cheap LM35 chips are not reliable and give fluctuating values. You may also have to use an RC filter circuit at the output of LM35 to increase reliability.

Program listing: Figure 6.10 shows the program listing (Program**: ONOFF**). At the beginning of the program, **SetTemp** is set to 20.0 Degrees, and the **LED** and **RELAY** are assigned to port numbers 3 and 2 respectively. Inside the **setup()** function, both the **RELAY** and **LED** are configured as outputs and are deactivated. Inside the main program loop, the room temperature is read and compared to the desired temperature. If the room temperature is lower than the desired temperature, then both the **LED** and **RELAY** are activated, otherwise they are deactivated. The program checks the temperature every 10 seconds. The **SetTemp** and **RoomTemp** are displayed on the LCD shown in Figure 6.11.

```
//---------------------------------------------------------------------
//                    ONOFF TEMPERATURE CONTROL
//                    =========================
//
// This is an ON-OFF temperature controller project. The ambient temperature
// is read and compared to the desired set value. If it is less than the set
// value then the relay and LED are activated, otherwise they are deactivated
//
// Author: Dogan Ibrahim
// File  : ONOFF
// Date  : June, 2023
//---------------------------------------------------------------------
#include <LCD_I2C.h>
LCD_I2C lcd(0x27, 16, 2);

float SetTemp = 20.0;                               // Desired temperature
#define LM35 A0                                     // LM35 at port A0
int raw;
int LED = 3;                                        // LED at port 3
int RELAY = 2;                                      // RELAY at port 2

void setup()
{
    pinMode(LED, OUTPUT);                           // LED is output
    pinMode(RELAY, OUTPUT);                         // RELAY is output
    digitalWrite(LED, LOW);                         // LED OFF at beginning
    digitalWrite(RELAY, LOW);                       // RELAY OFF at beginning
    lcd.begin();                                    // Initialize LCD
    lcd.backlight();                                // BAcklight ON
}

void loop()
{
  raw = analogRead(LM35);                           // Read temperature
  float mV = 5000.0 * raw / 1023.0;                 // in mV
  float RoomTemp = mV / 10.0;                       // Room temperature
```

```
  lcd.clear();                                   // Clear LCD
  lcd.setCursor(0, 0);
  lcd.print(" SetTemp = ");                      // Display SetTemp =
  lcd.print(SetTemp);                            // Display SetTEmp
  lcd.setCursor(0, 1);
  lcd.print("RoomTemp = ");                      // Display RoomTemp =
  lcd.print(RoomTemp);                           // Display RoomTemp

  if(SetTemp > RoomTemp)                         // If cold
  {
    digitalWrite(LED, HIGH);                     // LED ON
    digitalWrite(RELAY, HIGH);                   // RELAY ON

  }
  else
  {
    digitalWrite(LED, LOW);                      // LED OFF
    digitalWrite(RELAY, LOW);                    // RELAY OFF
  }
  delay(10000);                                  // Wait 10 seconds
}
```

Figure 6.10: Program: ***ONOFF****.*

Figure 6.11: The LCD display.

Suggestion: In Figure 6.10, the **SetTemp** is a single value. As a result of this, the relay may have to operate many times as the temperature fluctuates around this value. In practical applications, it is better to choose two values close to each other and then keep the temperature between these two values.

6.5 Project 4: Darkness reminder – using a light-dependent resistor (LDR)

Description: The LDR was introduced in Project 4 (Section 5.8) when you developed the conveyor belt goods counter project. In that project, LDR was used with one of the digital inputs of the development board.

The analog output voltage of the LDR can be used to calculate the level of brightness in a room. Alternatively, it can be used as an ON-OFF switch to detect when it becomes dark (i.e. dark sensor). The LDR can be used for example with a relay and motor to close the curtains at night time, or to switch ON the outside lights at night time, and so on.

In this project, LDR is used to detect when it becomes dark. The relay supplied with the kit is activated when darkness is detected, and also, the on-board LED is turned ON. The contacts of the relay can be connected to various devices which need to be activated when it becomes dark.

Block diagram: Figure 6.12 shows the block diagram of the project.

Figure 6.12: Block diagram of the project.

Circuit diagram: As described in Chapter 5, LDRs are simple resistors whose resistance decreases with increasing incident light. These devices are usually used in light-level control applications. Figure 6.13 shows the circuit diagram of the project. One leg of the LDR is connected to ground (0V). The other leg is connected to +5 V through a potentiometer. The junction of the LDR with the potentiometer is connected to analog input A0 of the development board. Notice that, just like resistors, LDRs have no polarities.

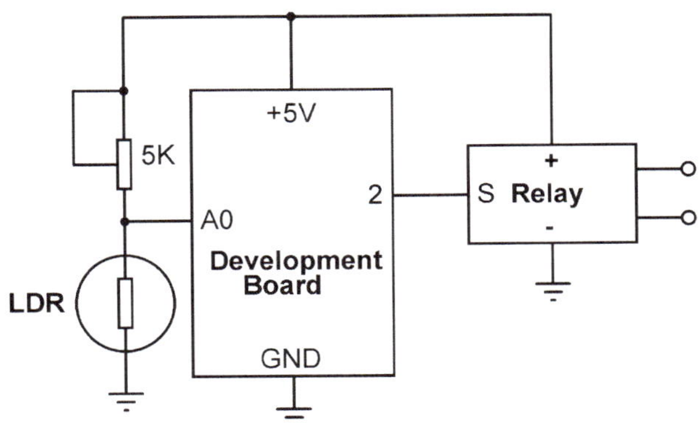

Figure 6.13: Circuit diagram of the project.

Program listing: Figure 6.14 shows the program listing (Program: **darkness**). At the beginning of the program, LDR is assigned to analog port A0. The relay is assigned to port 2 and is configured as output and is deactivated at the beginning of the program. Also, the on-board LED at port 13 is turned OFF. Inside the program loop, the output of the LDR is read and stored in integer variable **ldr**. In this project, darkness is assumed if **ldr** is greater than 800 (you will have to experiment by displaying the values of variable **ldr** using the Serial Monitor at different light levels. Note the value of variable **ldr** when there is no light falling on the LDR) and when this happens both the relay and the on-board LED are turned ON to indicate darkness. You can adjust the dark level detection point by varying the potentiometer,

```
//--------------------------------------------------------------------
//                    DARKNESS REMINDER
//                    =================
//
// In this project a LDR is usd with a relay. The relay and the on-board
// LED are turned ON when dark (i.e. when the light falling on the LDR
// is reduced)
//
// Author: Dogan Ibrahim
// File  : darkness
// Date  : June, 2023
//--------------------------------------------------------------------
#define LDR A0                                           // LDR at port A0
int ldr;
int LED = 13;                                            // On-board LED at 13
int RELAY = 2;                                           // RELAY at port 2

void setup()
{
    pinMode(LED, OUTPUT);                                // LED is output
    pinMode(RELAY, OUTPUT);                              // RELAY is output
    digitalWrite(LED, LOW);                              // LED OFF at beginning
    digitalWrite(RELAY, LOW);                            // RELAY OFF at beginning
}

void loop()
{
  ldr = analogRead(LDR);                                 // Read temperature
  if(ldr > 800)                                          // If dark detected
  {
    digitalWrite(LED, HIGH);                             // LED ON
    digitalWrite(RELAY, HIGH);                           // RELAY ON
  }
  else
  {
    digitalWrite(LED, LOW);                              // LED OFF
    digitalWrite(RELAY, LOW);                            // RELAY OFF
  }
}
```

Figure 6.14: Program: **darkness.**

6.6 Project 5: Tilt detection

Description: There are applications where you may want to know if an object is tilted. In this project, you will be using the supplied vibration tilt sensor device and activate the relay when the device is tilted. In normal applications the tilt sensor is attached to an object where you wish to detect when the object is tilted.

Vibration tilt sensor

Two SW-520D-type vibration tilt sensor devices (Figure 6.15) are supplied with the kit. This sensor consists of two metal balls that act as a switch. When the device is horizontal and the sensor angle is less than 10°, the switch is in the open state. When the device is tilted by more than 10°, the switch is closed.

Figure 6.15: SW-520D vibration tilt sensor.

Block diagram: Figure 6.16 shows the block diagram of the project.

Figure 6.16: Block diagram of the project.

Circuit diagram: The circuit diagram is shown in Figure 6.17. One leg of the sensor is connected to GND, while the other leg is connected to port 2. A 10-kΩ pull-up resistor is used to +5 V. The relay is connected to port 3.

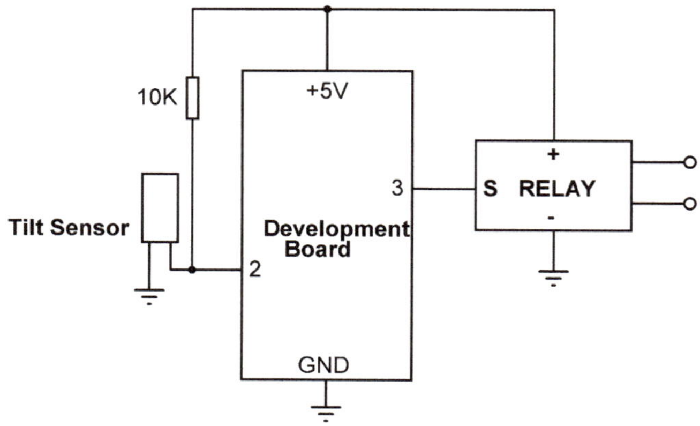

Figure 6.17: Circuit diagram of the project.

Program listing: Figure 6.18 shows the program listing (Program: **Tilt**). At the beginning of the program, **TILT** and **RELAY** are assigned to ports 2 and 3 respectively. **TILT** is configured as an input and **RELAY** as an output. Inside the main program loop, the state of the sensor is checked. If the sensor output is LOW (i.e., sensor is tilted and contacts are closed) then the relay is activated, otherwise, the relay is deactivated.

```
//----------------------------------------------------------------------
//                    TILT DETECTOR
//                    =============
//
// This is a tilt detector project. A tilt sensor is attache dto an object.
// If the object is tilted then the relay is activated
//
// Author: Dogan Ibrahim
// File  : Tilt
// Date  : June, 2023
//----------------------------------------------------------------------
int TILT = 2;                                    // Tilt at port 2
int RELAY = 3;                                   // RELAY at port 3

void setup()
{
    pinMode(TILT, INPUT);                        // TILT is input
    pinMode(RELAY, OUTPUT);                      // RELAY is output
    digitalWrite(RELAY, LOW);                    // RELAY OFF at beginning
}

void loop()
{
  int Sensor = digitalRead(TILT);                // Read the sensor state
  if(Sensor == LOW)                              // Sensor tilted
```

```
    digitalWrite(RELAY, HIGH);                      // Activate relay
  else
    digitalWrite(RELAY, LOW);                       // Deactivate relay
}
```

Figure 6.18: Program: **Tilt**.

6.7 Water level sensor

The water level sensor (Figure 6.19) is used to detect the presence of water, for example, the level of water in a tank, the level of water in a pool, to detect rainfall, to detect water leakage, and so on. The sensor has a series of ten exposed copper traces, five of which are power traces and five are sense traces. These traces are interlaced so that there is one sense trace between every two power traces. The traces are bridged by water when submerged in water or when water is present on the sensor. A small LED is lit on the sensor when power is applied.

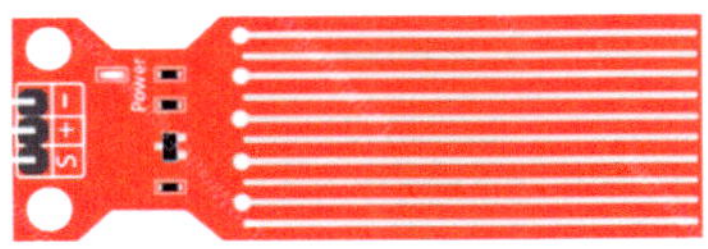

Figure 6.19: Water level sensor.

The sensor has 3 pins: **-** (GND), **+** (power supply), and **S** (analog output). the basic specifications of the sensor are:

- Operating voltage: 3.3 V to 5 V
- Current: less than 20 mA

One problem with the water level sensors is that the lifetime of these sensors is shortened when the sensor is exposed to a moist environment, especially when power is applied to the sensor. One way around this problem is not to power the sensor permanently, but to power it only when reading will be taken. This can be done if the sensor is powered from a digital output port of the development board since the output ports can supply the required 20 mA.

The output of the water level sensor is 0 V when it is not exposed to water. The output voltage increases as the sensor is submerged in water. The higher the water level, the higher will be the output voltage. Therefore, by measuring the output voltage of the sensor you can determine the level of water for example in a tank. Notice that the output voltage depends on the amount of minerals present in the water. It is therefore necessary to calibrate the sensor for the type of water you are using. This is explained in the next project.

6.7.1 Project 6: Displaying water level

Description: In this project, you will be using the water level sensor supplied with the kit. The sensor is connected to analog input A0 of the development board. The water level readings are displayed on the Serial Monitor when the sensor is inserted into a container with water as the level of water is increased (Figure 6.20). By knowing the relationship between the water level and the ADC output, you can use the sensor in water level control applications.

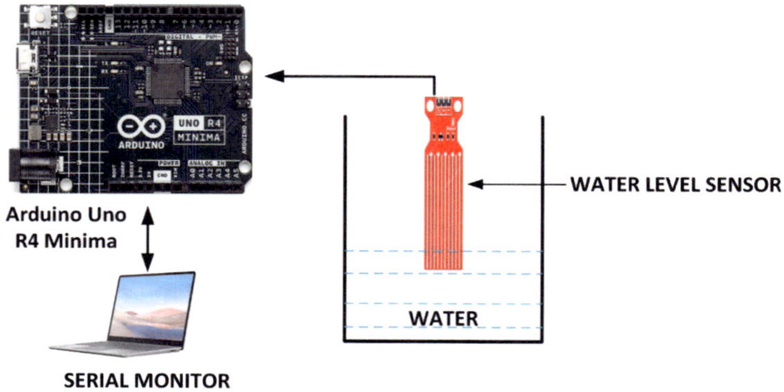

Figure 6.20: Measuring water level.

Circuit diagram: Figure 6.21 shows the circuit diagram of the project. The sensor is powered from pin 2 and its output (pin S) is connected to analog input A0 of the development board.

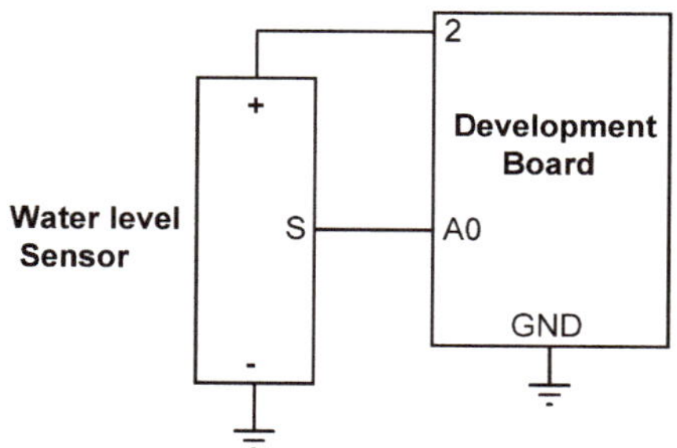

Figure 6.21: Circuit diagram of the project.

Program listing: Figure 6.22 shows the program listing (Program: **Waterlevel**). At the beginning of the program, pin 2 is assigned to **POWER** and is configured as output. This pin provides power to the sensor. The output of the sensor is connected to analog input A0 of the development board. Serial Monitor is used to display the values read by the ADC as more water is added to the cup. As shown in Table 6.1, a list is made showing the ADC output against the level of liquid in cm. This table can be used in other water sensor applications as long as the conditions are the same (e.g. same cup), e.g. for controlling the level of water in a tank.

Water level (cm)	ADC output
0	0
1	450
2	500
2.5	520
3	530
3.5	540
4	550

Table 6.1: Water level vs. ADC output.

```
//----------------------------------------------------------------------
//                    WATER LEVEL SENSOR
//                    ==================
//
// In this project a water level sensor is used. The program measures
// the water leproject measures the water level. Serial Monitor is used
// to tabulate the results
//
// Author: Dogan Ibrahim
// File  : Waterlevel
// Date  : June, 2023
//----------------------------------------------------------------------
int POWER = 2;                                   // Power at port 2
#define Sensor A0

void setup()
{
    pinMode(POWER, OUTPUT);                      // POWER is output
    Serial.begin(9600);                          // Serial monitor
    delay(5000);
}

void loop()
{
  digitalWrite(POWER, HIGH);                     // Apply power to sensor
  int raw = analogRead(Sensor);                  // Read sensor value
  Serial.println(raw);                           // Display ADC value
  delay(2000);                                   // One second delay
}
```

Figure 6.22: Program: **Waterlevel.**

6.7.2 Project 7: Water level controller

Description: In this project, the aim is to control the amount of water in a tank between two levels. It is assumed that Table 6.1 is valid in this project. i.e., the same type of water and the same tank are used as in the previous project.

Block diagram: Figure 6.23 shows the block diagram of the project. A relay-driven pump draws water from a reservoir and fills a tank. A water level sensor detects the amount of water in the tank. The pump (i.e. relay) is operated if the water level is below **LOWL** and is stopped when the water level in the tank reaches **HIGHL**. In this project, LOWL and HIGHL are set as 450 and 540 respectively (i.e., the water height should always be between 1 cm and 3.5 cm).

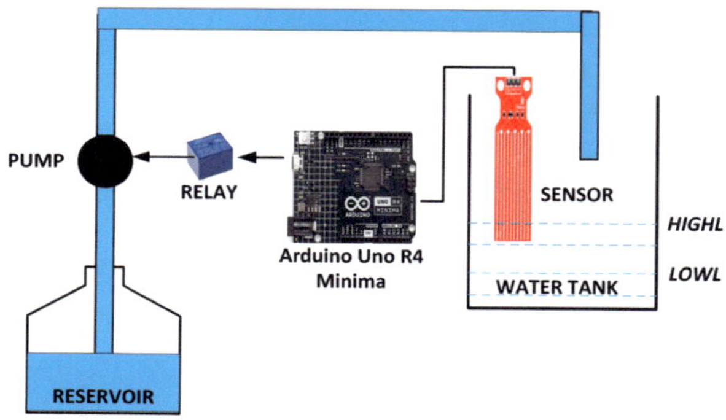

Figure 6.23: Block diagram of the project.

Circuit diagram: Figure 6.24 shows the circuit diagram. The sensor is powered from port 2 and sensor output is connected to analog input A0 as in the previous project. The relay is connected to port 3 of the development board.

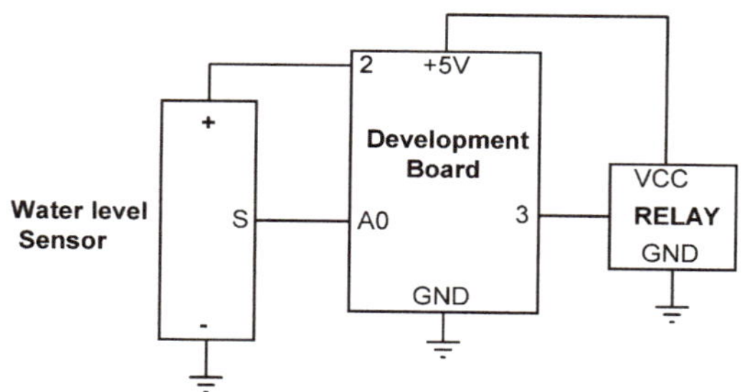

Figure 6.24: Circuit diagram of the project.

Program listing: Figure 6.25 shows the program listing (Program: **WaterControl**). Inside the **setup()** function, **POWER** and **RELAY** pins are configured as outputs and the relay is deactivated. Inside the main program loop, the water level is measured. If it is below **LOWL** then the relay is activated until the water reaches **HIGHL**. At this point, the relay is deactivated. Power is also removed from the sensor for 5 seconds where the measurement starts again.

```
//--------------------------------------------------------------------
//                    WATER LEVEL CONTROL
//                    ===================
//
// In this project a water level sensor is used. The program controls the
// amount of water in a tank between two levels LOWL and HIGHL. A relay
// is controlled to keep the water between these levels
//
// Author: Dogan Ibrahim
// File  : WaterControl
// Date  : June, 2023
//--------------------------------------------------------------------
int POWER = 2;                                         // Power at port 2
int RELAY = 3;                                         // RELAY at port 3
#define Sensor A0                                      // Sensor output
int LOWL = 450;                                        // Low level
int HIGHL = 540;                                       // High level

void setup()
{
    pinMode(POWER, OUTPUT);                            // POWER is output
    pinMode(RELAY, OUTPUT);
    digitalWrite(RELAY, LOW);
}

void loop()
{
  digitalWrite(POWER, HIGH);                           // Apply power to sensor
  int raw = analogRead(Sensor);                        // Read sensor value
  if(raw < LOWL)
  {
    digitalWrite(RELAY, HIGH);
    while(raw < HIGHL)
    {
      raw = analogRead(Sensor);                        // Read sensor value
    }
    digitalWrite(RELAY, LOW);
  }
  digitalWrite(POWER, LOW);                            // Remove power
  delay(5000);                                         // Five seconds delay
}
```

Figure 6.25: Program: ***WaterControl****.*

6.7.3 Project 8: Water flooding detector with buzzer

Description: In this project, the water sensor is placed near a water supply to detect water leakage or flooding. If water leakage has been detected, then a buzzer sounds as a warning. The buzzer could be replaced with a relay so that other more powerful warning devices can be activated.

Block diagram: Figure 6.26 shows the block diagram of the project.

Figure 6.26: Block diagram of the project.

Circuit diagram: The circuit diagram is shown in Figure 6.27. The water sensor is connected to the development board as in the previous project. The buzzer is connected to port 3.

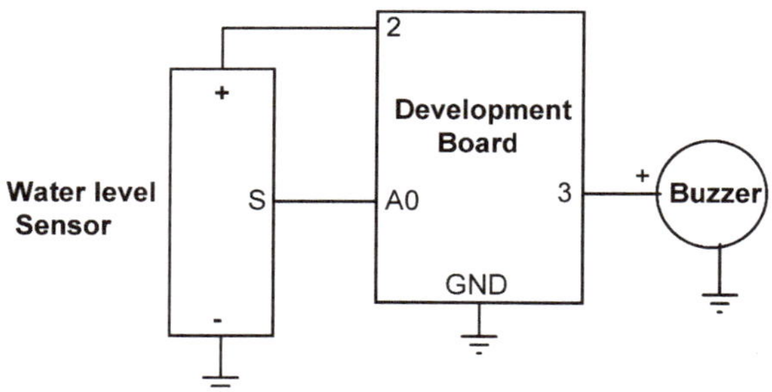

Figure 6.27: Circuit diagram of the project.

Program listing: Figure 6.28 shows the program listing (Program: **Flooding**). The trigger point is set to 500 in this project. Notice that once flooding is detected the buzzer keeps sounding until the processor is reset.

```
//-----------------------------------------------------------------------
//                    FLOODING WARNING
//                    ================
//
// In this project a water level sensor is used. The program detects
// water leakage/flooding and sounds the buzzer. The system must be
// reset to stop the buzzer sounding
//
```

```
// Author: Dogan Ibrahim
// File  : Flooding
// Date  : June, 2023
//----------------------------------------------------------------------
int POWER = 2;                                       // Power at port 2
int BUZZER = 3;                                      // RELAY at port 3
#define Sensor A0                                    // Sensor output
int Trigger = 500;                                   // Trigger level

void setup()
{
    pinMode(POWER, OUTPUT);                          // POWER is output
    pinMode(BUZZER, OUTPUT);                         // Buzzer is output
    digitalWrite(BUZZER, LOW);                       // Buzzer OFF
    digitalWrite(POWER, HIGH);                       // Apply power to sensor
}

void loop()
{
  int raw = analogRead(Sensor);                      // Read sensor value
  if(raw > Trigger)
  {
    digitalWrite(BUZZER, HIGH);                      // Buzzer ON
    while(1);                                        // Wait here forever
  }
}
```

Figure 6.28: Program: **Flooding**.

6.8 Project 9: Sound detection sensor — control the relay by clapping hands

Description: This project uses the sound detection sensor to toggle the state of a relay when you clap your hands close to the sensor.

The sound detection sensor

This is a small module (Figure 6.29) incorporating a sensitive capacitive microphone for detecting sound and an amplifier. The output of the module can be analog or digital. The sound sensitivity of the module is adjusted by the on-board potentiometer. The analog output voltage changes with the intensity of sound received by the microphone. You can connect this output to an analog input pin and process the output voltage. In this project, you will be using the digital output. The digital output goes LOW when sound with enough loudness is detected by the microphone.

The module has the following pins:

A0	analog output
G	GND
+	power supply
D0	digital output

Figure 6.29: Sound detection sensor.

Block diagram: Figure 6.30 shows the block diagram of the project.

Figure 6.30: Block diagram of the project.

Circuit diagram: The circuit diagram is shown in Figure 6.31. Digital output of the sensor module is connected to port 2 and the relay is connected to port 3.

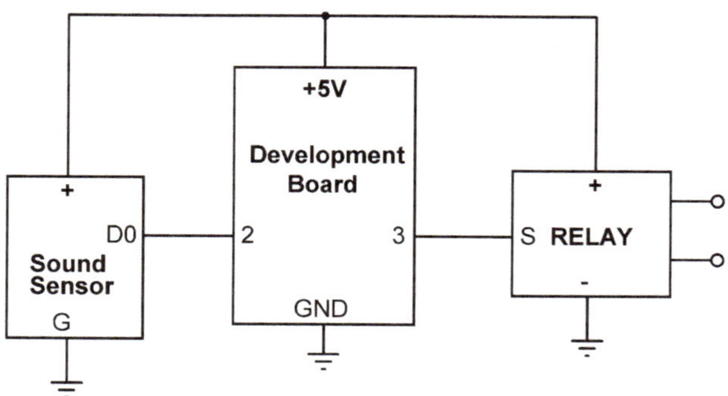

Figure 6.31: Circuit diagram of the project.

Program listing: Figure 6.32 shows the program listing (Program: **Clap**). The sensor output is assigned to port 2 and **RELAY** to port 3. The sensor is the input and **RELAY** is the output. Inside the main program, the output of the sensor is checked. When the sensor is activated by clapping next to the microphone, its output goes LOW. This is detected by the program which changes the state of the relay. You will have to adjust the sensitivity via the on-board potentiometer and LED.

```
//----------------------------------------------------------------------
//                  SOUND CONTROLLED RELAY
//                  ======================
//
// In this project the sound sensor module is used together with the
// relay. Clapping hands near the sensor toggles the state of the relay
//
// Author: Dogan Ibrahim
// File  : Clap
// Date  : June, 2023
//----------------------------------------------------------------------
int D0pin = 2;                                  // Sensor output port 2
int RELAY = 3;                                  // RELAY at port 3
bool RelayState = false;

void setup()
{
    pinMode(D0pin, INPUT);                      // D0 pin is input
    pinMode(RELAY, OUTPUT);                     // Relay is output
    digitalWrite(RELAY, LOW);                   // Relay OFF
}

void loop()
{
  while(digitalRead(D0pin) == 1);               // Wait for clap
  RelayState = !RelayState;                     // Change state
  if(RelayState)
      digitalWrite(RELAY, HIGH);                // Relay ON
  else
      digitalWrite(RELAY, LOW);                 // Relay OFF
}
```

Figure 6.32: Program: **Clap.**

6.9 Project 10: Flame sensor — fire detection with relay output

Description: In this project, the supplied flame sensor is used to detect fire and then activate the relay. The relay for example can be connected to a sound device to warn of fire.

The flame sensor

The supplied flame sensor (Figure 6.33) is basically a YG-1006 type 2-pin photosensitive transistor, sensitive to Infra-Red light wavelengths between 760 nm to 1100 nm which is the wavelength of flame. The device has two pins: Emitter (long pin) and Collector (short pin). The device is normally used in series with a 10-kΩ resistor, where the Collector is connected to +5 V and Emitter to GND through the resistor. Output is taken from the emitter-resistor junction. The phototransistor starts to conduct when it gets IR light from the flame. A comparator can be used to convert the output to a digital signal.

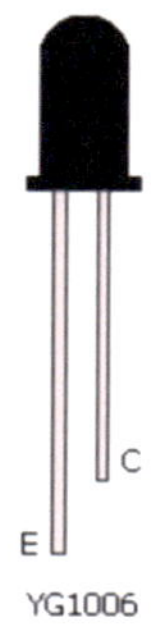

Figure 6.33: Supplied flame sensor.

Block diagram: Figure 6.34 shows the block diagram of the project.

Figure 6.34: Block diagram of the project.

Circuit diagram: Figure 6.35 shows the circuit diagram. The relay is connected to port 3 and output from the flame sensor is connected to analog input A0.

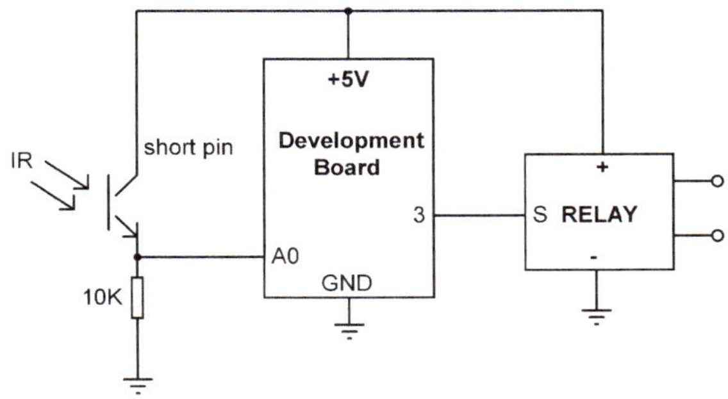

Figure 6.35: Circuit diagram of the project.

Program listing: Before using the project, you should run the Serial Monitor and display the ADC data output from the sensor with and without flame near the sensor. The author noticed that without any flame the sensor output was about 30. When there was flame from a match about a couple of feet away from the sensor, the reading went to over 500. In this project, the **Trigger** point is taken as 200.

Figure 6.36 shows the program listing (Program: **Flame**). RELAY is assigned to port 3 and configured as output. Inside the main program loop, the output of the flame sensor is read continuously and if it is above the **Trigger** value then it is assumed that fire has occurred and the relay is activated. The only way to deactivate the relay is to reset the processor.

```
//----------------------------------------------------------------------
//                    SOUND CONTROLLED RELAY
//                    ======================
//
// In this project the sound sensor module is used together with the
// relay. Clapping hands near the sensor toggles the state of the relay
//
// Author: Dogan Ibrahim
// File  : Flame
// Date  : June, 2023
//----------------------------------------------------------------------
#define FLAME A0
int RELAY = 3;                                  // RELAY at port 3
int Trigger = 200;

void setup()
{
    pinMode(RELAY, OUTPUT);                     // Relay is output
    digitalWrite(RELAY, LOW);                   // Relay OFF
}

void loop()
```

```
{
  int raw = analogRead(FLAME);                    // Read flame sensor
  if(raw > Trigger)                               // If flame detected
  {
    digitalWrite(RELAY, HIGH);                    // Relay ON
    while(1);                                     // Wait here forever
  }
}
```

Figure 6.36: Program: **Flame.**

6.10 Project 11: Temperature and humidity display

Description: In this project, the DHT11 temperature and relative humidity sensor chip is used to get the ambient temperature and the relative humidity. The readings are displayed every 5 seconds on the LCD. The aim of this project is to show how the popular DHT11 relative humidity and temperature sensor chip can be used in projects.

Block diagram: Figure 6.37 shows the block diagram of the project.

Figure 6.37: Block diagram of the project.

Circuit Diagram: The standard DHT11 is a 4-pin digital output device (only 3 pins are used) shown in Figure 6.38, having pins +V, GND, and Data. The Data pin is internally pulled up to +V through a 10-kΩ resistor. The chip uses a capacitive humidity sensor and a thermistor to measure the ambient temperature. Data output is available from the chip around every few seconds. The basic features of DHT11 are:

- 3 to 5 V operation
- 2.5 mA current consumption (during a conversion)
- Temperature reading in the range 0-50 °C with an accuracy of ±2 °C
- Humidity reading in the range of 20-80% with 5% accuracy
- Breadboard compatible with 0.1-inch pin spacings

Figure 6.38: The DHT11 module.

Figure 6.39 shows the circuit diagram of the project. Here, the Data output of the DHT11 is connected to pin 2 of the development board.

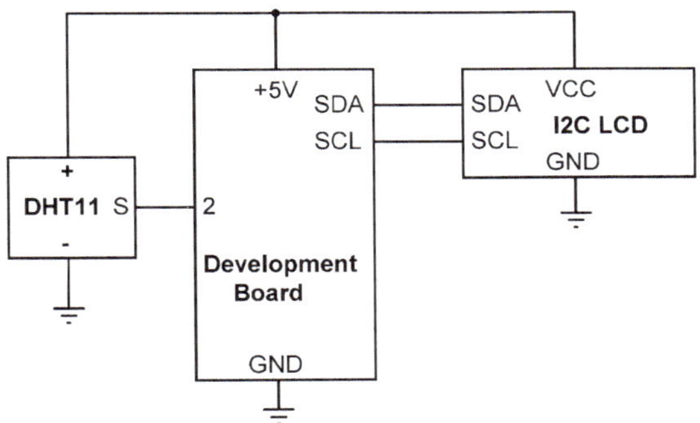

Figure 6.39: Circuit diagram of the project.

Program listing: It is necessary to install the DHT sensor library before the sensor chip can be used. The steps to install this library are given below (Figure 6.40):

- Start the IDE.
- Click to open the **LIBRARY MANAGER**.
- Type **dht sensor** in the search box and scroll down.
- Click **INSTALL** to install the **DHT sensor library by Adafruit**.
- Exit from the **LIBRARY MANAGER**.

At the time of writing this book the version of the library was 1.4.4

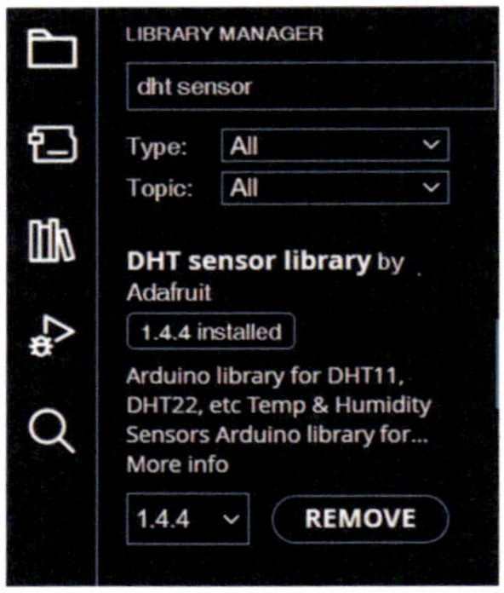

Figure 6.40: Install the Adafruit DHT library.

Program listing: Figure 6.41 shows the program listing (**DHT11monitor**). At the beginning of the program, the DHT header is included and the variables used in the program are declared. DHT11 is started in the **setup() function**. Function **ReadDHT11()** reads the temperature and humidity data from the sensor. You should notice that sometimes DHT11 does not return any data. This is detected by checking whether the returned data is not a number (using the keyword **isnan**). This function returns 1 to the main program is the data is read successfully, otherwise, 0 is returned. The main program displays the temperature and humidity on the LCD every 5 seconds. If the data is not read correctly, then the program attempts to read again after 5 seconds of delay.

```
//-----------------------------------------------------------------------
//            DHT11 TEMPERATURE AND HUMIDITY DISPLAY
//            ======================================
//
// In this project a DHT11 temperature and humidity sensor chip is used.
// The data read is displayed on the LCD every 5 seconds
//
// Author: Dogan Ibrahim
// File  : DHT11monitor
// Date  : June, 2023
//-----------------------------------------------------------------------
#include "DHT.h"
#include <LCD_I2C.h>
LCD_I2C lcd(0x27, 16, 2);

float hum, temp;
#define Sensor 2                                // DHT11 Sensor at pin 2
#define DHTTYPE DHT11                           // DHT11 is used
DHT dht(Sensor, DHTTYPE);

void setup()
{
  lcd.begin();                                  // initialize the lcd
```

```
  lcd.backlight();                            // Backlight ON
  dht.begin();
}

//
// Read the temperature and humidity. DHT11 sometimes fails to return
// data. Catch this condition and return correct status to teh calling
// program. Returning 1 is success, 0 is error
//
int ReadDHT11()
{
  hum = dht.readHumidity();                   // Read humidity
  temp = dht.readTemperature();               // Read temperature in C

  // Check if any read failed and return status to caller to try again
  if (isnan(hum) || isnan(temp))
  {
    return 0;
  }
  else
    return 1;
}


void loop()
{
  delay(2000);                                // DHT11 is a slow device
  if(ReadDHT11() == 1)                        // If successful read
  {
    lcd.clear();                              // Clear LCD
    lcd.setCursor(0, 0);                      // Cursor at top
    lcd.print("T=");                          // Display T=
    lcd.print(temp);                          // Display temperature
    lcd.print(" C");                          // Display C
    lcd.setCursor(0, 1);                      // Cursor at bottom
    lcd.print("H=");                          // Display H=
    lcd.print(hum);                           // Display humidity
    lcd.print(" %");                          // Display %
    delay(5000);                              // Wait 5 seconds
  }
  else                                        // Failed to read
    delay(3000);                              // Wait 3 seconds
}
```

Figure 6.41: Program: **DHT11monitor**.

Figure 6.42 shows an example display.

Figure 6.42: Example display.

6.11 Project 12: Generating musical tones — melody maker

Description: In this project, musical notes are generated to play the well-known melody **Happy Birthday**. The sound is sent to the piezo sounder supplied with the kit.

Block diagram: Figure 6.43 shows the block diagram of the project.

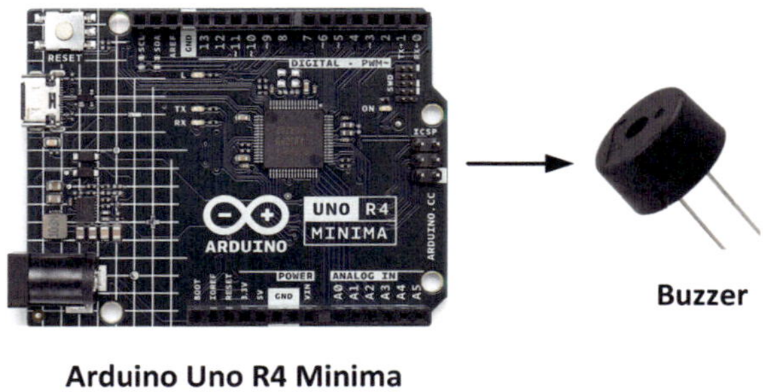

Figure 6.43: Block diagram of the project.

Circuit diagram: The piezo sounder is connected to pin 2 of the development board shown in Figure 6.44.

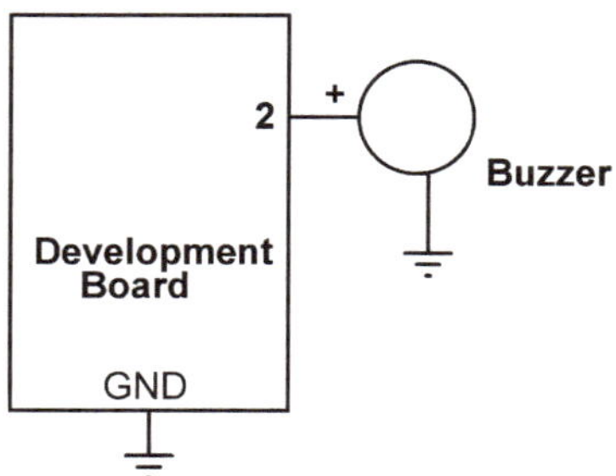

Figure 6.44: Circuit diagram of the project.

Program listing: When playing a melody each note is played for a certain duration and with a certain frequency. In addition, a certain gap is necessary between two successive notes. The frequencies of the musical notes starting from middle C (i.e. C4) are given below. The harmonic of a note is obtained by doubling the frequency. For example, the frequency of C5 is 2 × 262 = 524 Hz.

Note	**C4**	**C4#**	**D4**	**D4#**	**E4**	**F4**	**F4#**	**G4**	**G4#**	**A4**	**A4#**	**B4**
Hz	261.63	277.18	293.66	311.13	329.63	349.23	370	392	415.3	440	466.16	493.88

In order to play the tune of a melody, you need to know its musical notes. Each note is played for a certain duration and there is a certain time gap between two successive notes. The next thing you want is to know how to generate a sound with the required frequency and duration. In this project, you will be generating the classic **Happy Birthday** melody and thus you need to know the notes and their durations. These are given in the table below where the durations are in units of 300 milliseconds (i.e., the values given in the table should be multiplied by 300 to give the actual durations in milliseconds).

Note	C4	C4	D4	C4	F4	E4	C4	C4	D4	C4	G4	F4	C4	C4	C5	A4	F4	E4	D4	A4#	A4#	A4	F4	G4	F4
Duration	1	1	2	2	2	3	1	1	2	2	2	3	1	1	2	2	2	2	2	1	1	2	2	2	4

The built-in function **tone()** generates a square wave of the specified frequency (and 50% duty cycle) on a pin. A duration can be specified, otherwise, the wave continues until a call to function **noTone()**. Only one tone can be generated at a time. If a tone is already playing on a different pin, the call to **tone()** will have no effect. If the tone is playing on the same pin, the call will set its frequency.

The program listing (program: **Melody**) is shown in Figure 6.45. The frequencies and durations of the melody are stored in two arrays called **frequency** and **duration** respectively. Before the main program loop, the durations of each tone are calculated and stored in array **Durations** so that the main program loop does not have to spend any time to do these calculations. Inside the program loop, the melody frequencies are generated with the required durations using built-in function **tone()**. Notice that the tone output is stopped by calling function **noTone()**. A small delay (100 ms) is introduced between each tone. The melody is repeated after a 3-second delay.

You can try higher harmonics of the notes for clearer sound by multiplying the frequency with an integer number. The sound quality of the piezo sounder is not good at all. A loudspeaker can be used with an audio amplifier for much better and clearer sound quality.

```
//----------------------------------------------------------------------
//                        PLAY A MELODY
//                        =============
//
// This program plays the well-known melody Happy Birthday
//
// Author: Dogan Ibrahim
// File  : Melody
// Date  : June, 2023
//----------------------------------------------------------------------
int piezo = 2;                              // Piezo at port 2
const int MaxNotes = 25;                    // MAx 25 notes
int Durations[MaxNotes];


//
```

```
// Melody frequencies
//
  unsigned int frequency[] = {262,262,294,262,392,349,262,262,294,262,
                              392,349,262,262,524,440,349,330,294,466,
                              466,440,349,392,349};
//
// Frequency durations
//
  int duration[] = {1,1,2,2,2,3,1,1,2,2,2,3,1,1,2,2,2,2,
                    2,1,1,2,2,2,3};

void setup()
{
  pinMode(piezo, OUTPUT);                     // Piezo is output
  for(int k = 0; k < MaxNotes; k++)           // Durations
    Durations[k] = 300 * duration[k];
}


void loop()
{
  for(int k = 0; k < MaxNotes; k++)
  {
      tone(piezo, frequency[k]);
      delay(Durations[k]);
      delay(100);                             // Wait
  }

  noTone(piezo);
  delay(3000);                                // Stop 3 seconds
}
```

Figure 6.45: Program: **Melody**.

Chapter 7 • The RFID Reader

7.1 Overview

RFID is the acronym for Radio Frequency Identification and qualifies devices used for security and tracking purposes. An RFID system includes a reader card and a tag. Both are included in the kit. RFID uses electromagnetic fields to transfer data over short distances. RFID systems are mainly used in security applications. For example, they can be used to open a door where the person having the right tag is allowed to open the door.

The RFID reader included in the kit is known as the RC522 module (Figure 7.1), with the following basic specifications:

- Operating frequency: 13.56 MHz
- Operating voltage: +3.3 V
- Operation with both SPI bus and I²C bus

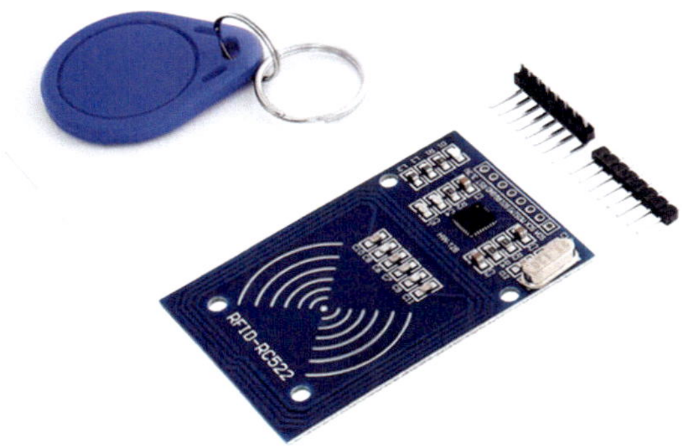

Figure 7.1: RFID reader and tag.

The RFID reader is supplied with header pins which must be soldered to the sockets at the edge of the reader before it can be used.

7.2 Project 1: Finding the Tag ID

Description: In this project, you will display the Tag ID of the supplied Tag on the Serial Monitor.

Block diagram: Figure 7.2 shows the block diagram of the project.

Figure 7.2: Block diagram of the project.

Circuit diagram: The connections between the development board ports and the RFID reader are as follows (*be careful not to connect the power pin to +5 V*). Figure 7.3 shows the circuit diagram of the project:

RFID reader pin	**Development board port**
SDA	10
SCK	13
MOSI	11
MISO	12
IRQ	not used
GND	GND
RST	9
3.3V	3.3 V

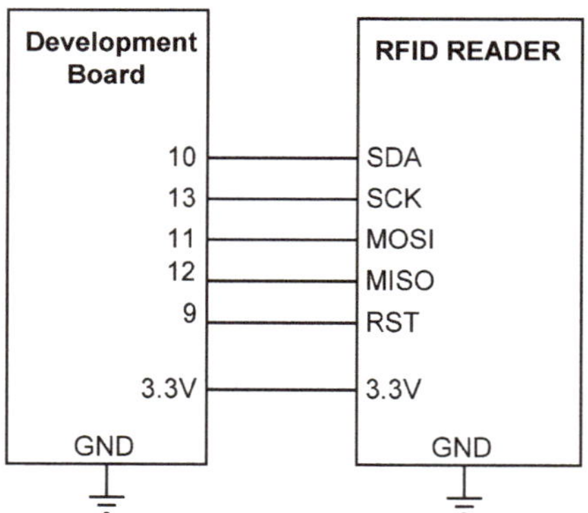

Figure 7.3: Circuit diagram of the project.

Program listing: Before using the RFID reader, you have to add the RFID library to your IDE. The library is named MFRC522 and the steps to add this library are as follows:

- Go to the following website and download the zip file **rfid-master.zip** to a folder: https://github.com/AritroMukherjee/RFID

- Start the IDE.
- Click **Sketch → Include Library → Add .zip Library**.
- Browse to the saved zip file and click **Open**.
- You can now start to use the library.

The RFID reader library offers many functions that can be seen by unzipping the library file. Some of the important library functions are:

mfrc522.PCD_Init()	Initialize the RFID reader
mfrc522.PICC_IsNewCardPresent())	Look for an RFID reader module
mfrc522.PICC_ReadCardSerial())	Select the RFID reader to use
mfrc522.uid.uidByte[]	Return the tag ID in an array
mfrc522.PICC_HaltA()	Stop reading (Halt PICC)

The program called **DumpInfo** given on the Arduino IDE website can be used to determine the Tag ID of your card. The steps are:

- Start the IDE.
- Click **File → Examples → MFRC522 → DumpInfo**.
- Compile and upload the program to the development board.
- Start the Serial Monitor.
- Place the white Tag on top of the reader and keep it there until the data display stops on the Serial Monitor.

You should now see data displayed similar to the one shown in Figure 7.4. This is the 1-kB memory data of the card and also its Tag ID. The 1-kB memory of the Tag is organized into 16 sectors (0 to 15), where each sector is further divided into 4 blocks (0 to 3). Each block can store 16 bytes of data (0 to 15).

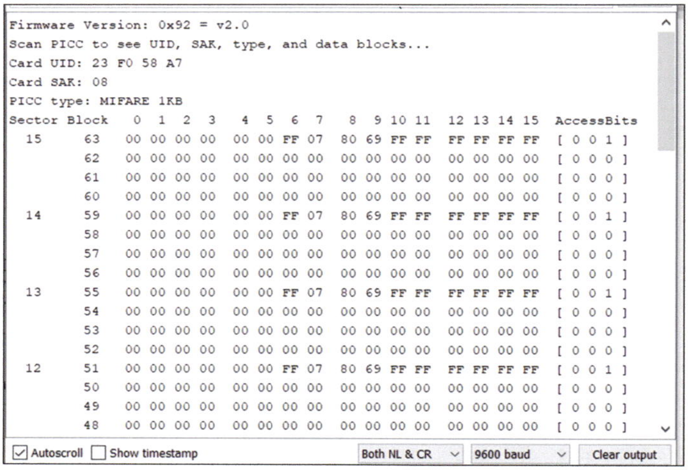

```
Firmware Version: 0x92 = v2.0
Scan PICC to see UID, SAK, type, and data blocks...
Card UID: 23 F0 58 A7
Card SAK: 08
PICC type: MIFARE 1KB
Sector Block   0  1  2  3   4  5  6  7   8  9 10 11  12 13 14 15  AccessBits
  15     63   00 00 00 00  00 00 FF 07  80 69 FF FF  FF FF FF FF  [ 0 0 1 ]
         62   00 00 00 00  00 00 00 00  00 00 00 00  00 00 00 00  [ 0 0 0 ]
         61   00 00 00 00  00 00 00 00  00 00 00 00  00 00 00 00  [ 0 0 0 ]
         60   00 00 00 00  00 00 00 00  00 00 00 00  00 00 00 00  [ 0 0 0 ]
  14     59   00 00 00 00  00 00 FF 07  80 69 FF FF  FF FF FF FF  [ 0 0 1 ]
         58   00 00 00 00  00 00 00 00  00 00 00 00  00 00 00 00  [ 0 0 0 ]
         57   00 00 00 00  00 00 00 00  00 00 00 00  00 00 00 00  [ 0 0 0 ]
         56   00 00 00 00  00 00 00 00  00 00 00 00  00 00 00 00  [ 0 0 0 ]
  13     55   00 00 00 00  00 00 FF 07  80 69 FF FF  FF FF FF FF  [ 0 0 1 ]
         54   00 00 00 00  00 00 00 00  00 00 00 00  00 00 00 00  [ 0 0 0 ]
         53   00 00 00 00  00 00 00 00  00 00 00 00  00 00 00 00  [ 0 0 0 ]
         52   00 00 00 00  00 00 00 00  00 00 00 00  00 00 00 00  [ 0 0 0 ]
  12     51   00 00 00 00  00 00 FF 07  80 69 FF FF  FF FF FF FF  [ 0 0 1 ]
         50   00 00 00 00  00 00 00 00  00 00 00 00  00 00 00 00  [ 0 0 0 ]
         49   00 00 00 00  00 00 00 00  00 00 00 00  00 00 00 00  [ 0 0 0 ]
         48   00 00 00 00  00 00 00 00  00 00 00 00  00 00 00 00  [ 0 0 0 ]
```


Figure 7.4: Tag memory data dump.

The 1-kB memory of the Tag is organized into 16 sectors (from 0 to 15). Each sector is further divided into 4 blocks (block 0 to 3). Each block can store 16 bytes of data (from 0 to 15). Therefore:

16 sectors × 4 blocks × 16 bytes = 1024 bytes of data on the card (i.e., 1 kB)

Block 3 of each sector (i.e. the top block) is called **Sector Trailer** and this contains the **Access Bits** which control the read/write access to the remaining blocks in the sector. Therefore, only the bottom 3 blocks (i.e. blocks 0, 1 and 2) of each sector are available for user data storage. This means that you actually have 48 bytes (3×16 bytes) per 64-byte sector for your own use.

Block 0 of sector 0 is known as the **Manufacturer Block/Manufacturer Data** and it contains the manufacturer data and the ID of the Tag.

The Tag ID is also displayed under the heading **Card UID** in Figure 7.4. In this project, your Tag ID is: **23 F0 58 A7**.

7.3 Project 2: RFID door lock access with relay

Description: In this project, the RFID reader and the supplied relay are both connected to the development board. It is assumed that a secure door entry is operated with a relay and is protected with an RFID system. Placing an authorized Tag card near the RFID reader can only open the door. The relay is activated for 15 seconds and after this time it is deactivated so that the door can be closed.

Block diagram: Figure 7.5 shows the block diagram of the project.

Figure 7.5: Block diagram of the project.

Circuit diagram: The circuit diagram is shown in Figure 7.6. The diagram is basically the same as in Figure 7.3, but here a relay is added to port 2.

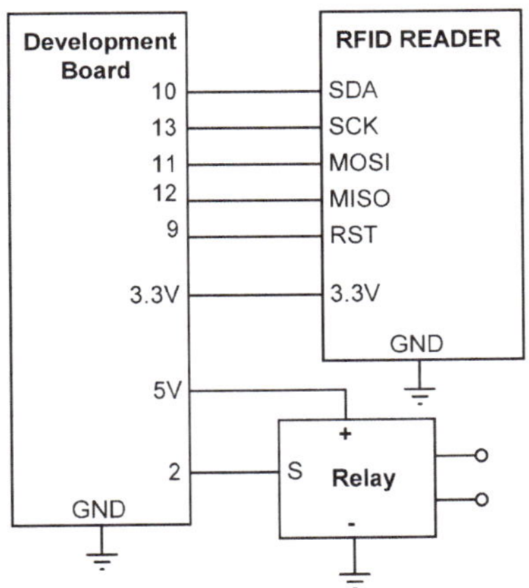

Figure 7.6: Circuit diagram of the project.

Program listing: Figure 7.7 shows the program listing (Program: **RFIDLock**). At the beginning of the program, the **SPI** and **MFRC522** libraries are included in the program. Valid card ID is stored in string **ValidCard**, **RELAY** is assigned to port 2. Inside the **setup()** function, **RELAY** is configured as output and is deactivated. **SPI** bus and **MFRC522** are also initialized. The remainder of the program runs inside the main program loop. Here, the program waits until a card is placed near the reader. When a card is placed, it is selected and a **for** loop is formed to read the 4-byte card Tag ID into variable **TagID**. The **TagID** is then compared with the authorized Tag ID **ValidCard**. If there is a match, the user is authorized and the relay is activated for 15 seconds. If the card is not valid, the relay remains deactivated. This process is repeated forever.

```
//-----------------------------------------------------------------------
//                  RFID LOCK SYSTEM
//                  ================
//
// In this program the RFID card reader is used with a relay. The relay
// is only activated if an authorized Tag is placed near the reader. The
// relay stays ON for 15 seconds and then turns OFF. The Tag ID of the
// authorized valid card in this example is: 23 F0 58 A7
//
// Author: Dogan Ibrahim
// File  : RFIDLock
// Date  : June, 2023
//-----------------------------------------------------------------------
#include <SPI.h>
#include <MFRC522.h>

#define SS_PIN 10
#define RST_PIN 9
MFRC522 mfrc522(SS_PIN, RST_PIN);           // Create MFRC522 inst
String ValidCard = "23F058A7";              // Valid Tag ID
String TagID = "";
int RELAY = 2;                              // RELAY at port 2
byte i;

void setup()
{
  pinMode(RELAY, OUTPUT);                   // RELAY is output
  digitalWrite(RELAY, LOW);                 // Deactivate RELAY
  SPI.begin();                              // Initiate SPI bus
  mfrc522.PCD_Init();                       // Initiate MFRC522
}

void loop()
{
  if (!mfrc522.PICC_IsNewCardPresent())     // Look for card
  {
    return;
  }

  if (!mfrc522.PICC_ReadCardSerial())       // Select the card
  {
    return;
  }

  TagID = "";
```

```
  for (i = 0; i < 4; i++)                    // Read 4 byte Tag ID
  {
    TagID.concat(String(mfrc522.uid.uidByte[i], HEX));
  }

  TagID.toUpperCase();                       // Convert to upper case
  mfrc522.PICC_HaltA();                      // Stop reading

  if(TagID == ValidCard)                     // Valid card?
  {
    digitalWrite(RELAY, HIGH);               // RELAY ON
    delay(15000);                            // Wait 15 seconds
    digitalWrite(RELAY, LOW);                // RELAY OFF
  }
  else
    digitalWrite(RELAY, LOW);                // RELAY OFF
}
```

Figure 7.7: Program: **RFIDLock***.*

Figure 7.8 shows the project built on the breadboard.

Figure 7.8: Project built on the breadboard.

Chapter 8 • The 4×4 Keypad

8.1 Overview

Keypads provide an uncomplicated way to let users interact with your projects. They can be used to enter passwords, control games and robots, navigate menus, and so on. In this chapter, you will develop projects on your development board using the supplied 4×4 pushbutton keypad.

The 4×4 Pushbutton keypad

The keypad supplied (Figure 8.1) with the kit is a 4×4 matrix, 16-button type with 4 row and 4 column connections. The keys are marked S1 through S16.

Figure 8.1: Supplied 4×4 keypad.

The structure of a 4×4 keypad is shown in Figure 8.1 (replace the key names with S1 through S16). There are 4 columns (C1, C2, C3, C4) and 4 rows (R1, R2, R3, R4).

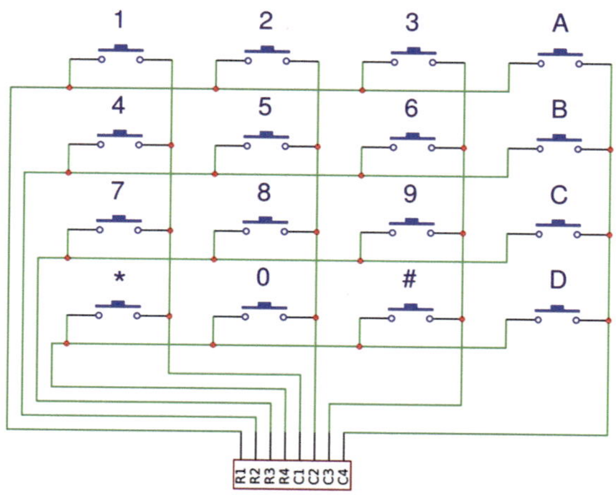

Figure 8.2: A 4×4 keypad structure.

The keypad works by a scanning process as follows:

- The row pins are connected to processor outputs, and column pins are connected to processor inputs with pullups so that the state of a pin is HIGH if the key is not pressed.
- The processor sets all row pins HIGH.
- The processor sets row 1 pins LOW.
- The processor reads the state of each column. If a column pin is HIGH then that button is not pressed. if a column pin is LOW, then that is the pressed key.
- The above process is repeated for the next row if a pressed button has not been detected.

A keypad library is provided for the Arduino IDE which makes simplifies the use of keypads in your projects. Some example projects are presented in the next sections.

8.2 Project 1: Display the pressed key code on the Serial Monitor

Description: This is a simple project where the pressed keys are displayed on the Serial Monitor. The aim of this project is to show how the keypad library can be used in projects.

Block diagram: Figure 8.3 shows the block diagram of the project.

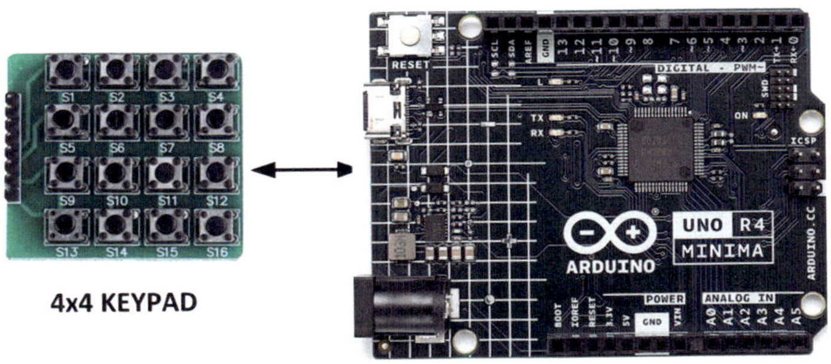

4x4 KEYPAD

Arduino Uno R4 Minima

Figure 8.3: Block diagram of the project.

Circuit diagram: The circuit diagram is shown in Figure 8.4. The connections between the keypad and the development board are as follows:

Keypad pin	**Development board port**
C4	2
C3	3
C2	4
C1	5
R1	6
R2	7
R3	8
R4	9

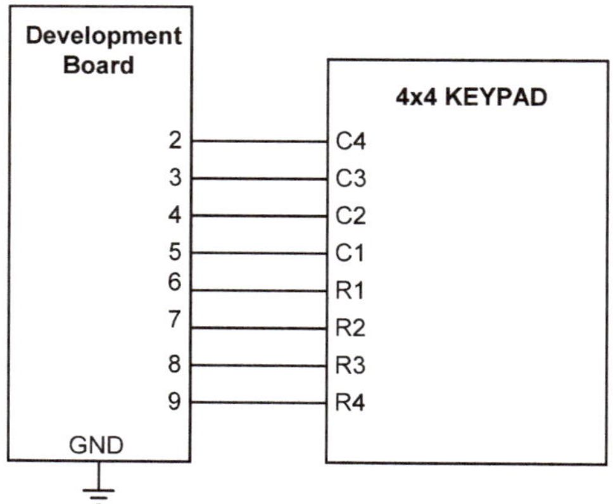

Figure 8.4: Circuit diagram of the project.

Program listing: At the time of drafting this book, only the **Adafruit Keypad library** was compatible with the Arduino Uno R4. Install this library as follows:

- Click to open the **LIBRARY MANAGER**.
- Search **Adafruit Keypad by Adafruit** in **LIBRARY MANAGER**.
- Click **INSTALL** to install the library.
- Exit the **LIBRARY MANAGER**.

At the time of authoring this book, the latest version of this library was 1.3.0.

Figure 8.5 shows the program listing (Program: **KeypadTest**). At the beginning of the program, the keypad library is included, and the number of rows and columns is specified. The array **keys** stores the button names in a 4×4 matrix. Notice that the buttons are labelled as **1** to **9** and then **A** to **G**. Then, the connections between the keypad and the development board ports are defined. Inside the main program, a key is read, and its value is displayed on the Serial Monitor.

```
//------------------------------------------------------------------------
//                    KEYPAD TEST PROGRAM
//                    ===================
//
// This program is used to test the keypad. The keys pressed on the
// keypad are displayed on the Serial Monitor
//
// Author: Dogan Ibrahim
// File  : KeypadTest
// Date  : June, 2023
//------------------------------------------------------------------------
```

```
#include "Adafruit_Keypad.h"

const byte ROWS = 4;
const byte COLS = 4;

char keys[ROWS][COLS] = {{'1', '2', '3', '4'},
                         {'5', '6', '7', '8'},
                         {'9', 'A', 'B', 'C'},
                         {'D', 'E', 'F', 'G'}};

byte rowPins[ROWS] = {9, 8, 7, 6};
byte colPins[COLS] = {2, 3, 4, 5};

Adafruit_Keypad MyKeys = Adafruit_Keypad(
makeKeymap(keys),rowPins,colPins,ROWS,COLS);

void setup()
{
  Serial.begin(9600);
  MyKeys.begin();
  delay(5000);
}

void loop()
{
  MyKeys.tick();

  while(MyKeys.available())
  {
    keypadEvent e = MyKeys.read();
    if(e.bit.EVENT == KEY_JUST_PRESSED) Serial.println((char)e.bit.KEY);
  }

}
```

Figure 8.5: Program: **KeypadTest**.

Figure 8.6 shows the display when keys 1 to G are pressed.

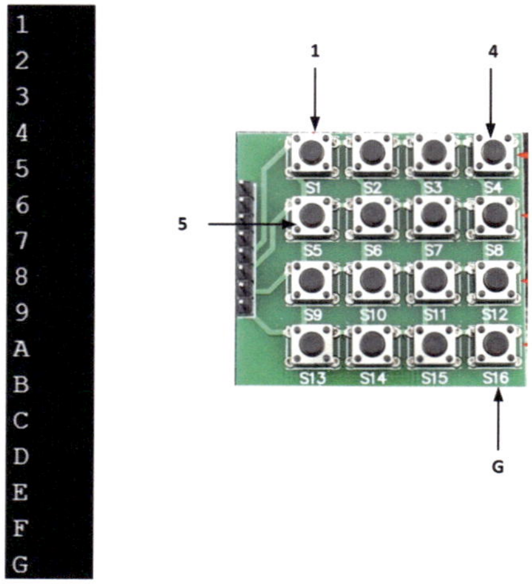

Figure 8.6: Display when keys are pressed.

8.3 Project 2: Integer calculator with LCD

Description: This is a simple integer calculator project. The calculator can perform the four basic operations: **+ – * /** on integer numbers. Results are displayed on the LCD.

The operation of the calculator is as follows: when power is applied to the system, the LCD displays the text **CALCULATOR** for 2 seconds. Then text **No1:** is displayed in the first row of the LCD and the user is expected to type the first number and then press the **ENTER (E)** key. Then text **No2:** is displayed in the second row of the LCD where the user enters the second number and press the **ENTER** key. After this, the required operation should be entered. The result will be displayed on the LCD for 5 seconds and then the LCD will be cleared, ready for the next calculation. The example below shows how numbers 12 and 20 can be added:

No1: 12 ENTER
No2: 20 ENTER
Op: + ENTER
12 + 20 = 32

The buttons are given the names as shown in Figure 8.7. The names used in the program for the buttons are as follows:

```
1 2 3 4
5 6 7 8
9 0   E
+ - * /
```

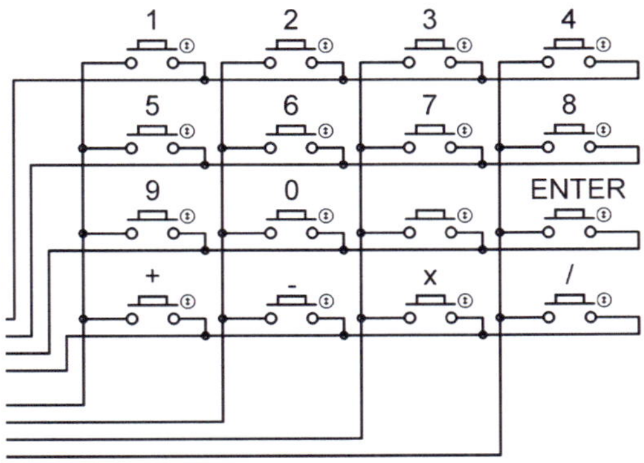

Figure 8.7: Keypad buttons.

Block diagram: Figure 8.8 shows the block diagram of the project.

Figure 8.8: Block diagram of the project.

Circuit diagram: The circuit diagram is shown in Figure 8.9. The keypad is connected as in the previous project. The LCD is connected as in the previous projects using the LCD.

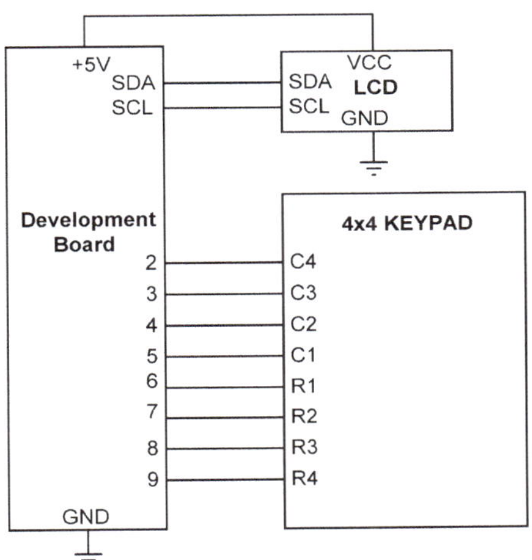

Figure 8.9: Circuit diagram of the project.

Program listing: Figure 8.10 shows the program listing (Program: **KeypadCalc**). At the beginning of the program, Keypad and I^{2}C LCD libraries are included and the keypad button layout is defined. Inside the **setup()** function, the LCD is initialized and the backlight is turned ON. The program then displays the text **CALCULATOR** for 2 seconds. After this, the user enters the first number with response to prompt **No1**:. Multi digit numbers can be entered. Then the second number is entered this time in response to **No2**:. Finally, the operation is entered. The Enter key (**E**) must be pressed after entering the numbers or the operation. A **switch** statement is used to determine what type of mathematical operation is required. The result is stored in variable **Calc** which is displayed on the LCD at the end of the program.

```
//------------------------------------------------------------------------
//                    INTEGER CALCULATOR
//                    ==================
//
// This is an integer calculator program using a keypad and LCD
// Basic operations of +-*/ can be performed
//
// Author: Dogan Ibrahim
// File  : KeypadCalc
// Date  : June, 2023
//------------------------------------------------------------------------
#include "Adafruit_Keypad.h"
#include <LCD_I2C.h>
LCD_I2C lcd(0x27, 16, 2);

const int ROWS = 4;                         //4 rows
const int COLS = 4;                         //4 columns
char key, ky;
unsigned long Calc, Op1, Op2;
unsigned char MyKey, Op;
unsigned int KeyNo;

char keys[ROWS][COLS] =                     //Button names
{
  {'1','2','3', '4'},
  {'5','6','7', '8'},
  {'9','0',' ', 'E'},
  {'+','-','*', '/'}
};

byte rowPins[ROWS] = {9, 8, 7, 6};
byte colPins[COLS] = {2, 3, 4, 5};

Adafruit_Keypad MyKeys = Adafruit_Keypad(
makeKeymap(keys),rowPins,colPins,ROWS,COLS);
```

```
void setup()
{
  lcd.begin();                          //Initialize LCD
  lcd.backlight();                      //Backlight ON
  MyKeys.begin();                       //Initialize Keypad
}

void loop()
{
  lcd.clear();                          // Clear LCD
  lcd.setCursor(0, 0);                  // Cursor at (0,0)
  lcd.print("CALCULATOR");              // Display heading
  delay(2000);                          // 2 seconds delay
  lcd.clear();                          // Clear display

//
// Get first number
//
  lcd.setCursor(0, 0);
  Op1 = 0;
  lcd.print("No1: ");                   // Display No1:
  while(1)
  {
    while(!MyKeys.available())
    {
      MyKeys.tick();
      keypadEvent e = MyKeys.read();
      if(e.bit.EVENT == KEY_JUST_PRESSED)
      {
        ky=(char)e.bit.KEY;
        break;
      }
    }

      if(ky == 'E')break;               // If E, exit
      KeyNo = ky - '0';
      lcd.print(KeyNo);                 // Display digits while entered
      Op1 = 10*Op1 + KeyNo;             // Total so far
  }

 //
 // Get second number
 //
   lcd.setCursor(0, 1);
   Op2 = 0;
```

```
  lcd.print("No2: ");                     // Display No2:
  while(1)
  {
    while(!MyKeys.available())
    {
      MyKeys.tick();
      keypadEvent e = MyKeys.read();
      if(e.bit.EVENT == KEY_JUST_PRESSED)
      {
        ky=(char)e.bit.KEY;
        break;
      }
    }

      if(ky == 'E')break;                 // If E, exit
      KeyNo = ky - '0';
      lcd.print(KeyNo);                   // Display digits while entered
      Op2 = 10*Op2 + KeyNo;               // Total so far
  }

//
// Get operation
//
  lcd.clear();
  lcd.setCursor(0, 0);
  lcd.print("Op: ");                      // Display Op:
  while(1)
  {
    while(!MyKeys.available())
    {
      MyKeys.tick();
      keypadEvent e = MyKeys.read();
      if(e.bit.EVENT == KEY_JUST_PRESSED)
      {
        ky=(char)e.bit.KEY;
        break;
      }
    }

      if(ky == 'E')break;                 // If E, exit
      Op = ky;
      lcd.print(char(Op));                // Display operation
  }

  switch(Op)
  {
    case '+':
```

```
        Calc = Op1 + Op2;                           // Addition required
        break;
      case '-':
        Calc = Op1 - Op2;                           // Subtraction required
        break;
      case '*':
        Calc = Op1 * Op2;                           // Multiplication required
        break;
      case '/':
        Calc = Op1 / Op2;                           // Division required
        break;
    }

    lcd.clear();                                    // Clear display
    lcd.setCursor(0, 0);
    lcd.print(Op1);                                 // Display first number
    lcd.print(char(Op));                            // Display operation
    lcd.print(Op2);                                 // Display second number
    lcd.print("=");                                 // Display =
    lcd.print(Calc);                                // Display result

    delay(5000);
    lcd.clear();
  }
```

Figure 8.10: Program: **KeypadCalc**.

Figure 8.11 shows the steps in carrying out a simple addition.

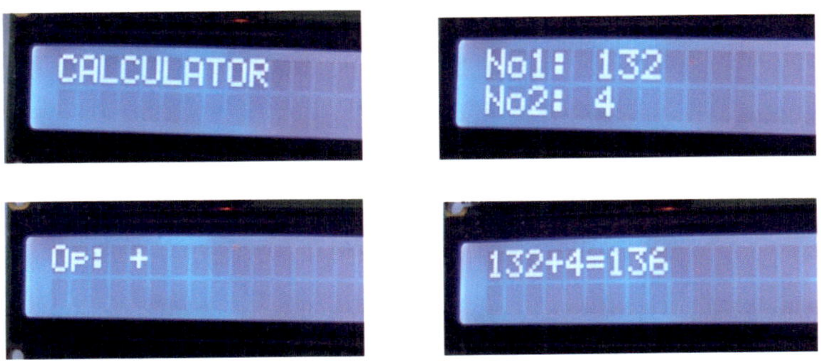

Figure 8.11: Steps in addition.

8.4 Project 3: Keypad door security lock with relay

Description: This is a keypad-based door security lock. It is assumed that the secure door is controlled with a relay such that activating the relay opens the door. In this project, the password is hardcoded as **1357** and the relay will only be activated if the correct code is entered on the keypad. The password must be terminated by pressing the Enter € key.

The relay will stay ON for 15 seconds and then it will be deactivated so that the door can be closed. You should press the **E** key before entering the secret key so that any previous numbers for example entered by mistake or entered by unauthorized people are cancelled.

Block diagram: Figure 8.12 shows the block diagram of the project.

Figure 8.12: Block diagram of the project.

Circuit diagram: The relay is connected to port 10 of the development board. The keypad is connected as in the previous project. Figure 8.13 shows the circuit diagram.

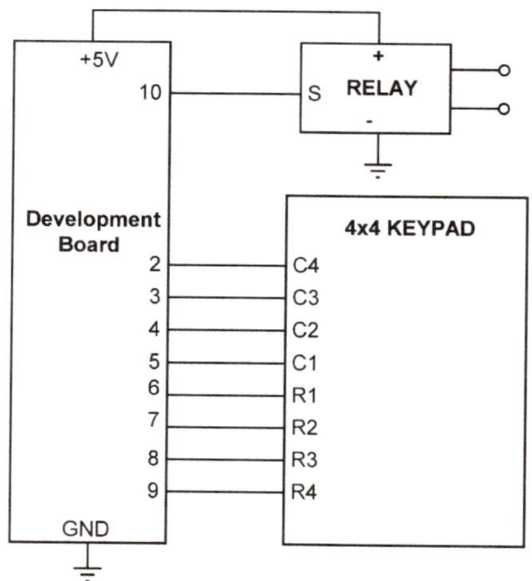

Figure 8.13: Circuit diagram of the project.

Program listing: Figure 8.14 shows the program listing (Program: **KeypadLock**). At the beginning of the program, the **KEYPAD** library is included, **SecretCode** is set to **1357**, and the keypad buttons are defined. Also, **RELAY** is assigned to port 10. Inside the **setup()** function, **RELAY** is configured as output and is deactivated. The program then reads the secret code from the keypad until the **E** key is pressed. The user-entered number is compared to the hardcoded secret code and if they are equal then the relay is activated for 15 seconds, otherwise, the relay remains OFF. Notice that you should enter the **E** key before entering the secret code so that any previous numbers entered, for example, by mistake, are cancelled.

```
//------------------------------------------------------------------
//                    KEYPAD LOCK WITH RELAY
//                    ======================
//
// This is a keypad lock project. A relay is activated if the
// correct code is entered on the keypad. Relay remains active
// for 15 seconds. The secret code is 1 3 5 7 follwed by E
//
// Author: Dogan Ibrahim
// File  : KeypadLock
// Date  : June, 2023
//------------------------------------------------------------------
#include "Adafruit_Keypad.h"

const int ROWS = 4;                              // 4 rows
const int COLS = 4;                              // 4 columns
unsigned char MyKey, ky;
unsigned int KeyNo;
unsigned long int SecretCode = 1357;             // Secret code
unsigned long int UserCode;                      // User entered code
int RELAY = 10;                                  // RELAY on port 10

char keys[ROWS][COLS] =                          // Keypad names
{
  {'1','2','3', '4'},
  {'5','6','7', '8'},
  {'9','0',' ', 'E'},
  {' ',' ',' ', ' '}
};

byte rowPins[ROWS] = {9, 8, 7, 6};
byte colPins[COLS] = {2, 3, 4, 5};

Adafruit_Keypad MyKeys = Adafruit_Keypad(
makeKeymap(keys),rowPins,colPins,ROWS,COLS);

void setup()
{
  pinMode(RELAY, OUTPUT);                        // RELAY is output
  digitalWrite(RELAY, LOW);                      // RELAY OFF
  MyKeys.begin();
}

void loop()
{
//
```

```
// Get the user code
//
  MyKey = 0;
  UserCode = 0;

  while(1)
  {
    while(!MyKeys.available())
    {
      MyKeys.tick();
      keypadEvent e = MyKeys.read();
      if(e.bit.EVENT == KEY_JUST_PRESSED)
      {
        ky = (char)e.bit.KEY;
        break;
      }
    }

      if(ky == 'E')break;                    // If E, exit
      KeyNo = ky - '0';
      UserCode = 10*UserCode + KeyNo;        // User code
  }

  if(UserCode == SecretCode)
  {
    digitalWrite(RELAY, HIGH);               // RELAY ON
    delay(15000);                            // 15 secs delay
    digitalWrite(RELAY, LOW);                // RELAY OFF
  }
  else
    digitalWrite(RELAY, LOW);
}
```

Figure 8.14: Program: **KeypadLock**.

Suggestion: you may like to choose a long secret code that's not easily guessed by trial and error. Here, you have 16 keys. For example, by choosing an 8-digit secret code, the number of permutations is 168 which is over billions assuming that the numbers can be repeated!

Chapter 9 • The Real-Time Clock (RTC) Module

9.1 Overview

The RTC module is an accurate clock module for use in microcontroller-based applications. The module provides seconds, minutes, hours, days, dates, months, and year information. The leap year is set automatically, where the module is reportedly valid until the year 2100.

9.2 The supplied RTC module

The supplied RTC module (Figure 9.1) is based on synchronous serial communication (not I^2C compatible). As shown in the image, it uses the DS1302 RTC chip with a quartz crystal for timing. A CR2032-coin type battery (not supplied due to transport restrictions) can be used to keep the clock running when the module is disconnected from the MCU.

Figure 9.1: The supplied RTC module.

The module has the following 4 pins:

VCC	power supply (+5 V)
GND	GND
CLK	clock
DAT	data
RST	reset

Two projects are given in this chapter. The first project is based on using the Serial Monitor to set the date/time and also to display the date/time and this is mainly for learning how to use the RTC module. The second project is based on using LCD to display the current date/time already set using the Serial Monitor.

9.3 Project 1: RTC with Serial Monitor

Description: In this project, you learn to set the current date/time using the Serial Monitor. The date and time will then be displayed on the Serial Monitor as the RTC is running.

Block diagram: Figure 9.2 shows the block diagram of the project.

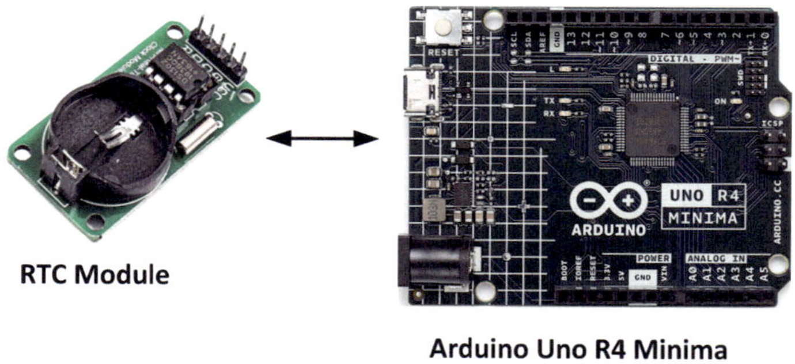

Figure 9.2: Block diagram of the project.

Circuit diagram: The connections between the development board and the RTC module are shown in Figure 9.3. The interface between the development board ports and the RTC module is as follows:

RTC module	**Development board port**
RST	2
DAT	3
CLK	4
GND	GND
VCC	+5V

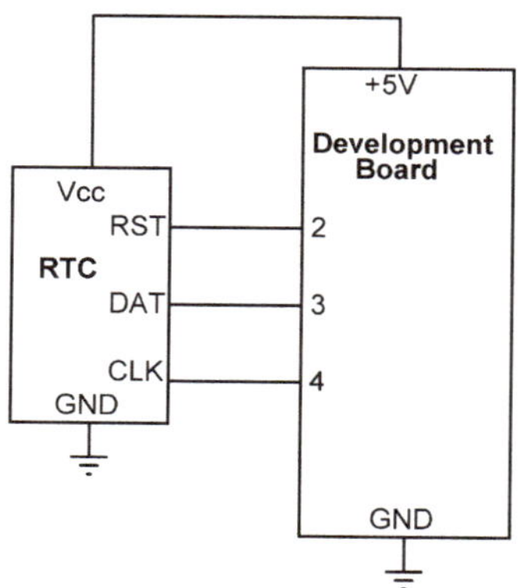

Figure 9.3: Circuit diagram of the project.

Program listing: There are several RTC libraries for the Arduino IDE. The one you will be using in this project is called **Virtuabotix** library. The steps to add this library to your IDE are as follows:

- Start the IDE.
- Clock **Sketch → Include Library → Add .ZIP Library**.
- Browse and select the **ArduinoRTCLibrary-master.zip** file.

Figure 9.4 shows the program listing (Program: **RTC**). At the beginning of the program, the RTC library is included. Inside the **setup()** function, the current date and time are set on the RTC using the following statement:

RTC.setDS1302Time(00, 52, 18, 4, 19, 7, 2023);

Where the parameters are:

Seconds:	00
Minutes:	52
Hours:	18
Day of week:	4
Day of month:	19
Month:	7
Year:	2023

```
//------------------------------------------------------------------------
//                    REAL TIME CLOCK MODULE
//                    ======================
//
// In this program we setup the RTC module at the current date and time
// using the Serial Monitor
//
// Author: Dogan Ibrahim
// File  : RTC
// Date  : June, 2023
//------------------------------------------------------------------------
#include <virtuabotixRTC.h>
#include <EEPROM.h>
virtuabotixRTC RTC(4, 3, 2);                    // CLK=4 ,DAT=3,RST=2

void setup()
{
  EEPROM.write(0,0);
  Serial.begin(9600);                          // Serial Monitor
  delay(5000);

  //
  // Set the current date and time as:
  // seconds,minutes,hours,day of the week,day of the month,month,year
  // If address 0 of EEPROM is 0xAA then it is assumed that date and time
  // are correct, otherwise new date and time must be entered
  //
  int chk = EEPROM.read(0);
  if(chk != 0xAA)
  {
```

```
      RTC.setDS1302Time(00, 52, 18, 5, 19, 7, 2023);
      EEPROM.write(0, 0xAA);                          // Write to EEPROM
      delay(100);
    }
  }

  //
  // Display leading 0 if the item is less than 10
  //
  void Display(int d)
  {
    if(d < 10)Serial.print("0");
    Serial.print(d);
  }


  void loop()
  {
    RTC.updateTime();                               // Update for display

    Serial.print("Current Date & Time: ");
    Display(RTC.dayofmonth);                        // Display day of month
    Serial.print("/");
    Display(RTC.month);                             // Display month
    Serial.print("/");
    Display(RTC.year);                              // Display year
    Serial.print(" ");
    Display(RTC.hours);                             // Display hours
    Serial.print(":");
    Display(RTC.minutes);                           // Display minutes
    Serial.print(":");
    Display(RTC.seconds);                           // Display seconds
    Serial.println();
    delay(1000);
  }
```

Figure 9.4: Program: **RTC**.

The above statement must be set once or when it is required to update the date or time (e.g. after a power loss), otherwise the current date and time may be wrong. In this program data 0xAA is written to the first location of the EEPROM after setting the date and time. The next time the program runs, it checks address 0 of the EEPROM and if that contains data 0xAA then it is assumed that both the data and time are correct and the above statement is not executed. The current date and time are displayed inside the main program loop. Function **Display()** displays a leading 0 if the value to be displayed is under 10. Figure 9.5 shows a sample display on the Serial Monitor.

```
Current Date & Time: 19/07/2023 18:54:38
Current Date & Time: 19/07/2023 18:54:39
Current Date & Time: 19/07/2023 18:54:40
Current Date & Time: 19/07/2023 18:54:41
Current Date & Time: 19/07/2023 18:54:42
Current Date & Time: 19/07/2023 18:54:43
Current Date & Time: 19/07/2023 18:54:44
Current Date & Time: 19/07/2023 18:54:45
Current Date & Time: 19/07/2023 18:54:46
Current Date & Time: 19/07/2023 18:54:47
```

Figure 9.5: Sample display produced by the program.

9.4 Project 2: RTC with LCD

Description: In this project, you will display the current date/time on the LCD.

Block diagram: Figure 9.6 shows the block diagram of the project.

Figure 9.6: Block diagram of the project.

Circuit diagram: The circuit diagram is shown in Figure 9.7. The RTC module and the LCD are connected to the development board as in the previous projects.

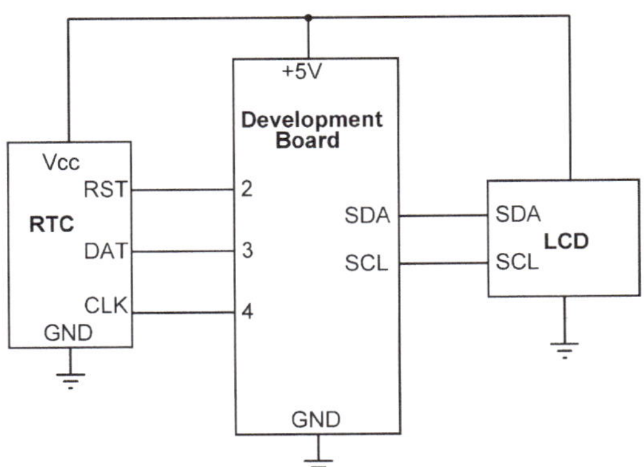

Figure 9.7: Circuit diagram of the project.

Program listing: Figure 9.8 shows the program listing (Program: **RTCLCD**). The program is very similar to the one in Figure 9.4, but here date and time are displayed on LCD. Figure 9.9 shows a sample display.

```
//-------------------------------------------------------------------
//                    REAL TIME CLOCK MODULE
//                    ======================
//
// In this program the RTC date and time are displayed on LCD
//
// Author: Dogan Ibrahim
// File  : RTCLCD
// Date  : June, 2023
//-------------------------------------------------------------------
#include <LCD_I2C.h>
#include <virtuabotixRTC.h>

virtuabotixRTC RTC(4, 3, 2);                       // CLK=4 ,DAT=3,RST=2
LCD_I2C lcd(0x27, 16, 2);

void setup()
{
  lcd.begin();                                     // Initialize the lcd
  lcd.backlight();                                 // Backlight ON
}

//
// Display leading 0 if the item is less than 10
//
void Display(int d)
{
  if(d < 10)lcd.print("0");
  lcd.print(d);
}


void loop()
{
  lcd.clear();
  RTC.updateTime();                                // Update for display
  lcd.setCursor(0, 0);

  Display(RTC.dayofmonth);                         // Display day of month
  lcd.print("/");
  Display(RTC.month);                              // Display month
  lcd.print("/");
  Display(RTC.year);                               // Display year
  lcd.print(" ");
  lcd.setCursor(0, 1);
  Display(RTC.hours);                              // Display hours
```

```
  lcd.print(":");
  Display(RTC.minutes);                    // Display minutes
  lcd.print(":");
  Display(RTC.seconds);                    // Display seconds
  delay(1000);
}
```

Figure 9.8: Program: **RTCLCD**.

Figure 9.9: Sample display.

9.5 Project 3: Temperature and humidity display with time stamping

Description: In this project, the ambient temperature and humidity are read and displayed on the Serial Monitor with time-stamping.

Block diagram: Figure 9.10 shows the block diagram of the project.

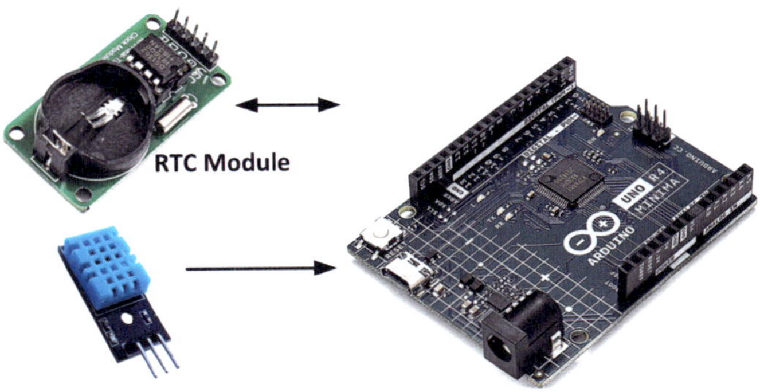

Figure 9.10: Block diagram of the project.

Circuit diagram: The circuit diagram is shown in Figure 9.11. The RTC module is connected to the development board as in the previous project.

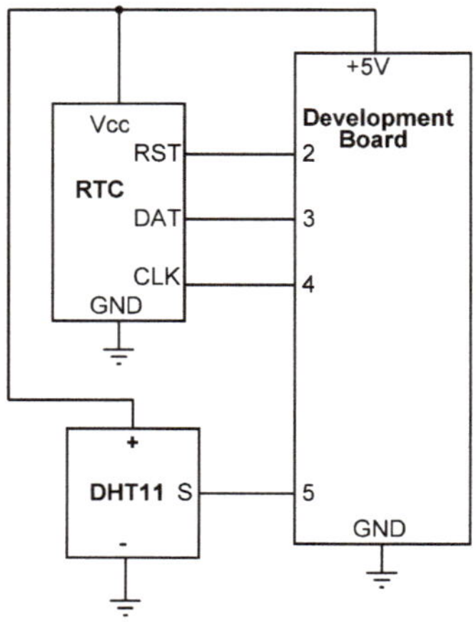

Figure 9.11: Circuit diagram of the project.

Program listing: Figure 9.12 shows the program listing (Program: **RTCDHT11**). At the beginning of the program, the RTC and the DHT11 libraries are included. Inside the **setup()**, DHT11 library is initialized and Serial Monitor is enabled. Function **ReadDHT11()** reads the temperature and humidity. Inside the main program loop, the current date and time are read and displayed together with the temperature and humidity data.

```
//------------------------------------------------------------------------
//                  TIME STAMPED TEMPERATURE AND HUMIDITY
//                  =====================================
//
// In this program the temperature and humidity readings are time stamped
// and displayed on the Serial Monitor every 5 seconds
//
// Author: Dogan Ibrahim
// File  : RTCDHT11
// Date  : June, 2023
//------------------------------------------------------------------------
#include "DHT.h"
#include <virtuabotixRTC.h>
virtuabotixRTC RTC(4, 3, 2);                     // CLK=4 ,DAT=3,RST=2
float hum, temp;
#define Sensor 5                                 // DHT11 at port 5
#define DHTTYPE DHT11                            // DHT11 is used
DHT dht(Sensor, DHTTYPE);

void setup()
{
   dht.begin();
   Serial.begin(9600);
   delay(5000);
```

```
}

//
// Read the DHT11 sensor temperature and humidity
//
int ReadDHT11()
{
  hum = dht.readHumidity();                    // Read humidity
  temp = dht.readTemperature();                // Read temperature in C

  // Check if any read failed and return status to caller to try again
  if (isnan(hum) || isnan(temp))
  {
    return 0;
  }
  else
    return 1;
}


//
// Display leading 0 if the item is less than 10
//
void Display(int d)
{
  if(d < 10)Serial.print("0");
  Serial.print(d);
}


void loop()
{
  if(ReadDHT11() == 1)                         // If successful read
  {
    RTC.updateTime();                          // Update for display
    Display(RTC.dayofmonth);                   // Display day of month
    Serial.print("/");
    Display(RTC.month);                        // Display month
    Serial.print("/");
    Display(RTC.year);                         // Display year
    Serial.print(" ");
    Display(RTC.hours);                        // Display hours
    Serial.print(":");
    Display(RTC.minutes);                      // Display minutes
    Serial.print(":");
    Display(RTC.seconds);                      // Display seconds
```

```
    Serial.print("  T=");                       // Display T=
    Serial.print(temp);                         // DIsplay temperature
    Serial.print("C");                          // Display C
    Serial.print("  H=");                       // Display H=
    Serial.print(hum);                          // Display humidity
    Serial.println("%");                        // Display %
  }
  delay(5000);                                  // Every 5 seconds
}
```

Figure 9.12: Program: **RTCDHT11**.

Figure 9.13 shows the sample output on the Serial Monitor.

```
19/07/2023 19:15:17  T=23.00C  H=35.00%
19/07/2023 19:15:22  T=23.00C  H=35.00%
19/07/2023 19:15:28  T=23.00C  H=35.00%
19/07/2023 19:15:33  T=23.00C  H=35.00%
19/07/2023 19:15:38  T=23.00C  H=35.00%
19/07/2023 19:15:43  T=23.00C  H=35.00%
```

Figure 9.13: Sample output.

9.6 Using the built-in RTC

The RTC on the UNO R4 Minima can be accessed using the RTC library that is included in the Renesas processor core. This library allows you to set/get the time as well as using alarms to trigger interrupts.

To set the starting time for the RTC, you can create an **RTCTime** object. Here you can specify the day, month, year, hour, minute, second, and specify the day of the week as well as daylight saving mode. Then to set the time, use the **setTime()** method. To retrieve the time, you need to use the **getTime()** method. The following methods can be used to get the date and time:

- getDayOfMonth()
- getMonth()
- getYear()
- getHour()
- getMinutes()
- getSeconds()

An example project is given below.

9.6.1 Project 4: Setting and displaying the current time

Description: In this project, the current time will be set and then it will be displayed every second on the Serial Monitor.

Program listing: Figure 9.14 shows the program listing (Program: **RenesasRTC**). The current date and time are set and then displayed every second as shown in Figure 9.15.

```
//-----------------------------------------------------------------------
//            SET THE CURRENT DATE AND TIME AND THEN DISLAY IT
//            ================================================
//
// This program uses the RTC on the Renesas processor. The current date
// and time are set and then displayed every second on Serial Monitor
//
// Author: Dogan Ibrahim
// File  : RenesasRTC
// Date  : June, 2023
//-----------------------------------------------------------------------
#include "RTC.h"

void setup()
{
   Serial.begin(9600);
   delay(5000);

   RTC.begin();
   RTCTime startTime(15, Month::JULY, 2023, 17, 37, 00, DayOfWeek::SATURDAY,
SaveLight::SAVING_TIME_ACTIVE);
   RTC.setTime(startTime);
}

void loop()
{
  RTCTime currentTime;
  RTC.getTime(currentTime);

  Serial.print(currentTime.getDayOfMonth());
  Serial.print("/");
  Serial.print(Month2int(currentTime.getMonth()));
  Serial.print("/");
  Serial.print(currentTime.getYear());
  Serial.print(" - ");
//
// Print time in format: HH:MM:SS
//
  Serial.print(currentTime.getHour());
  Serial.print(":");
  Serial.print(currentTime.getMinutes());
  Serial.print(":");
  Serial.println(currentTime.getSeconds());
```

```
  delay(1000);
}
```

Figure 9.14: Program: **RenesasRTC***.*

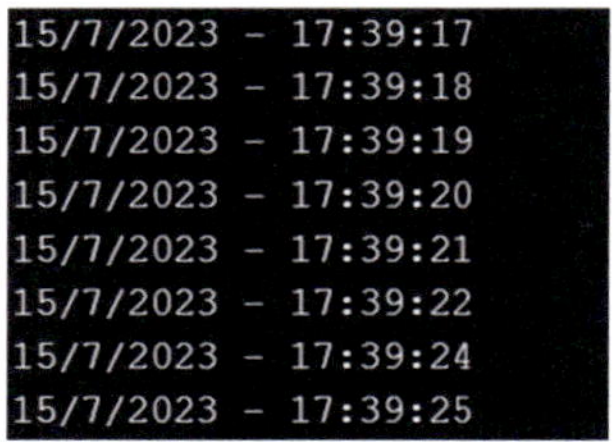

Figure 9.15: Displaying on the Serial Monitor.

The RTC can be used to set periodic interrupts. An example project is given below.

9.6.2 Project 5: Periodic interrupt every 2 seconds

Description: In this project, the on-board LED will be flashed every 2 seconds using RTC-based periodic interrupts.

Program listing: Figure 9.16 shows the program listing (Program: **RTCLED**). Inside the setup() function, the on-board LED is configured as output. It is important that the **RTC. setTime()** must be called for the periodic interrupts to work. It does not matter what the date and time are set to. Function **ActLED** is configured as the callback function which is called automatically every 2 seconds. Inside this function, the LED is toggled.

The periodic interrupts support the following periods:

ONCE_EVERY_2_SEC
ONCE_EVERY_1_SEC
N2_TIMES_EVERY_SEC
N4_TIMES_EVERY_SEC
N8_TIMES_EVERY_SEC
N16_TIMES_EVERY_SEC
N32_TIMES_EVERY_SEC
N64_TIMES_EVERY_SEC
N128_TIMES_EVERY_SEC
N256_TIMES_EVERY_SE

```
//------------------------------------------------------------------------
//          PERIODIC INTERRUPTS
//          ===================
//
// This program uses the RTC on the Renesas processor to generate
// periodic interrupts at every 2 seconds.The on-board LED is flashed
```

```
// when a periodic interrupt occurs (i.e. every 2 seconds)
//
// Author: Dogan Ibrahim
// File  : RTCLED
// Date  : July, 2023
//----------------------------------------------------------------------
#include "RTC.h"

const int LED = 13;                             // On-board LED

void ActLED()                                   // Callback function
{
  static bool flag = false;

  if(flag == true)
    digitalWrite(LED, HIGH);
  else
    digitalWrite(LED, LOW);

  flag = !flag;
}

//
// Configure the periodic interrupts. RTC.setTime() must be called for the
periodic
// interrupts to work. It does not matter what the date and time are set to
//
void setup()
{
   pinMode(LED, OUTPUT);

   RTC.begin();
   RTCTime mytime(25, Month::AUGUST, 2022, 14, 37, 00, DayOfWeek::THURSDAY,
SaveLight::SAVING_TIME_ACTIVE);
   RTC.setTime(mytime);
   RTC.setPeriodicCallback(ActLED, Period::ONCE_EVERY_2_SEC);
}

void loop()
{
}
```

Figure 9.16: Program: ***RTCLED****.*

Note: The Arduino Uno R4 WiFi board has a VRTC pin for connecting an external battery (1.6 – 3.6 V) permitting the on-board RTC to run even when the board is not powered. On the Uno R4 Minima board, there is no such pin, and the RTC is powered from the +5 V supply rail.

Chapter 10 • The Joystick

10.1 Overview

Joysticks are used in many game consoles, in robotics, in mouse control applications, and so on. In this chapter, you will learn the basic operation of the joystick supplied with the kit and discover how you can use it.

10.2 The joystick

The joystick supplied with the kit (Figure 10.1) is an analog device that is similar to two potentiometers connected together, one for the horizontal movement (X-axis) and one for the vertical movement (Y-axis). Additionally, the joystick includes a pushbutton switch. The main aim of a joystick is to communicate motion in 2D. As the joystick arm is moved left-right or up-down, the resistances of the two potentiometers change, and this data can be used to detect the movement in 2D.

Figure 10.1: The supplied joystick.

The joystick is connected to the external world with a set of pins. The main output pins are VRX and VRY which determine how far the joystick arm is pushed in the X or Y dimensions:

GND	GND
+5	power supply
VRX	Joystick reading in the horizontal direction (X-axis)
VRY	joystick reading in the vertical direction (Y-axis)
SW	normally open (HIGH) switch output. Pushing the SW switch turns its output LOW

10.3 Project 1 — Reading analog values from the joystick

Description: In this project, you will connect your joystick to the development board's analog ports and display the outputs on the Serial Monitor as the joystick arm is moved.

Block diagram: Figure 10.2 shows the block diagram of the project.

Figure 10.2: Block diagram of the project.

Circuit diagram: The circuit diagram is shown in Figure 10.3. The VRX and VRY pins are connected to analog inputs A0 and A1, respectively. Switch SW is connected to port 2.

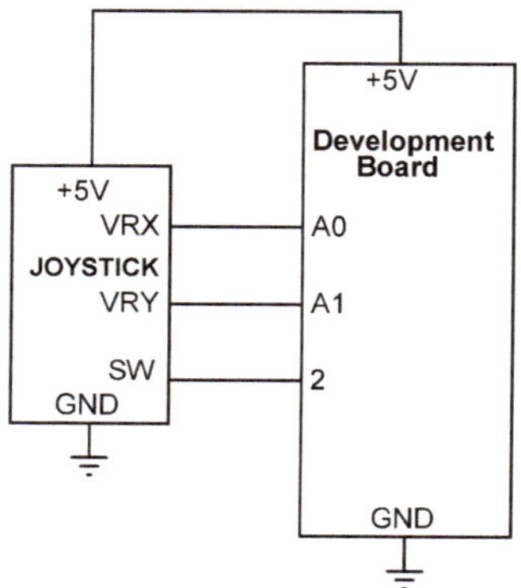

Figure 10.3: Circuit diagram of the project.

Program listing: Figure 10.4 shows the program listing (Program: **Joystick**). In this program, the ADC is used in default 10-bit mode. At the beginning of the program, the connections between the joystick and the development board ports are defined. Inside the **setup()** function, the Serial Monitor is initialized to 9600 baud (bit/s). Inside the main program loop, you read the analog X and Y values as the joystick arm is moved left, right, up, and down. You also read the switch output and display it on the Serial Monitor.

Figure 10.5 shows an example display of the Serial Monitor.

```
//----------------------------------------------------------------------
//                    JOYSTICK
//                    ========
//
// In this program the movements of the joystick are displayed
//
// Author: Dogan Ibrahim
// File  : Joystick
// Date  : July, 2023
```

```
//-----------------------------------------------------------------
#define X A0                                // X axis
#define Y A1                                // Y axis
int SW = 2;                                 // SW pin

void setup()
{
  pinMode(SW, INPUT_PULLUP);                // SW pin is input
  Serial.begin(9600);                       // Serial Monitor
  delay(5000);
}


void loop()
{
  int Xdir = analogRead(X);                 // Read X values
  int Ydir = analogRead(Y);                 // Read Y values

  Serial.print(Xdir);                       // Display X values
  Serial.print("\t");                       // Tab
  Serial.print(Ydir);                       // Display Y values
  Serial.print("\t");                       // Tab
  Serial.println(digitalRead(SW));          // Display switch value
  delay(250);
}
```

Figure 10.4: Program: ***Joystick****.*

504	508	1
504	508	1
504	508	1
504	508	1
504	508	1
504	508	1
504	508	1
504	508	1
504	508	1
504	508	1
503	508	1
504	508	1
504	508	1
504	508	1
504	508	1
504	508	1
504	508	1
504	508	1

Figure 10.5: Example display.

Observations (Connector of joystick at the left):

Table 10.1 shows the results obtained by the author when the joystick arm is moved, as well as when the switch is pressed.

Arm position	A0 reading (X-axis)	A1 reading (Y-axis)	SW reading
Idle	504	508	1
Fully right	1019	508	1
Fully left	0	508	1
Fully up	504	0	1
Fully down	504	1019	1
SW pressed arm idle	504	508	0

Table 10.1: Results when the joystick arm is moved.

The default ADC of the Arduino is 10 bits, which corresponds to 1024 levels. When the joystick arm is idle, you expect the readings to be 512 at both axes. You obtained 504 and 508 in the idle position, which are very close to the theoretical value. Also, when fully in one direction you'd expect 1023. The value 1019 obtained is also very close to the theoretical value.

Mapping the output

In some applications, you may want to map the joystick output of 0 to 1023 to some other range, say, 0 to 100, so that 100, for example, corresponds to full movement in one direction, and 50 corresponds to the idle position. This is done using the **map** statement as shown in the program in Figure 10.6 (Program: **Joystick2**).

```
//----------------------------------------------------------------------
//                          JOYSTICK
//                          ========
//
// In this program the movements of the joystick are displayed
// In this version of the program the output is mapped 0 to 100
//
// Author: Dogan Ibrahim
// File  : Joystick2
// Date  : July, 2023
//----------------------------------------------------------------------
#define X A0                                  // X axis
#define Y A1                                  // Y axis
int SW = 2;                                   // SW pin

void setup()
{
  pinMode(SW, INPUT_PULLUP);                  // SW pin is input
  Serial.begin(9600);                         // Serial Monitor
```

```
  delay(5000);
}


void loop()
{
  int Xdir = analogRead(X);                  // Read X values
  int Ydir = analogRead(Y);                  // Read Y values

  int mapinX = map(Xdir,0, 1023, 0, 100);
  int mapinY = map(Ydir, 0, 1023, 0, 100);

  Serial.print(mapinX);                      // Display X values
  Serial.print("\t");                        // Tab
  Serial.print(mapinY);                      // Display Y values
  Serial.print("\t");                        // Tab
  Serial.println(digitalRead(SW));           // Display switch value
  delay(250);
}
```

Figure 10.6: Program: ***Joystick2****.*

Figure 10.7 shows the output after the mapping.

```
99      49      1
99      49      1
99      49      1
99      49      1
99      49      1
99      49      1
99      49      1
49      49      1
49      49      1
49      49      1
49      49      1
49      49      1
49      49      1
49      49      1
49      49      1
```

Figure 10.7: Output after mapping.

Chapter 11 • The 8×8 LED Matrix

11.1 Overview

LED matrix devices are used in many graphical applications. In this chapter, you learn to use the 8×8 LED matrix included in the kit.

11.2 The supplied 8×8 LED matrix

The supplied 8×8 LED matrix (Figure 11.1) is a type 1588BS with 8 rows and 8 columns, i.e., 64 LEDs. This is a red 1.5-inch display with 3.7-mm dot size and a current draw of 5 to 20 mA. The voltage drop of each LED is 1.8 to 2.0 V.

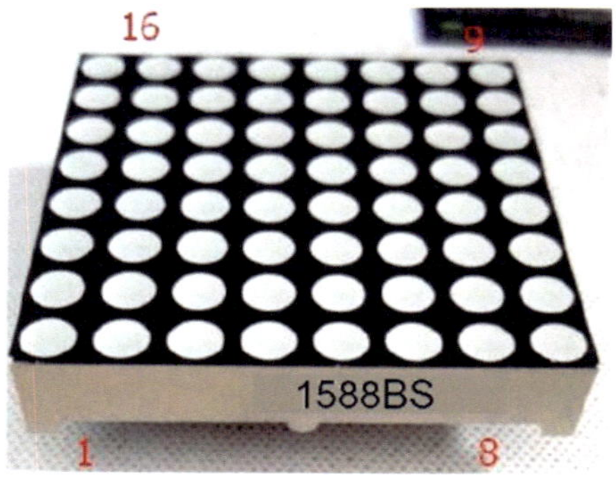

Figure 11.1: Supplied LED matrix.

The display has 16 pins, where the pin layout is shown in Figure 11.2. Pin 1 is located at the bottom left where the display is type-labelled. The bottom pins are 1 to 8, where pin 8 starts from top right side and goes up to 16 at the top left.

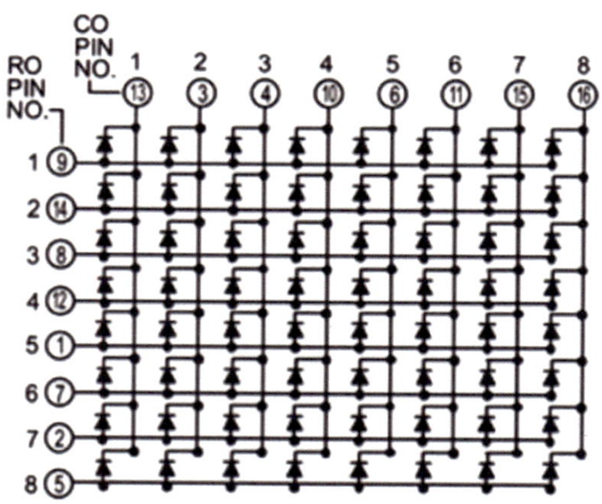

Figure 11.2: Pin layout of the display.

The display is normally controlled using 16 pins. This too much though and would occupy almost all available pins of the Arduino Uno. In this chapter, you start to use the 74HC595 shift register chip to reduce the number of pins required to 11 (8 for the LEDs and 3 for control).

11.3 Project 1: Displaying shapes

Description: In this project, you will display an UP ARROW shape on the 8×8 LED matrix. The aim of the project is to show how the 74HC595 shift register can be used to display shapes on the small LED matrix.

Block diagram: Figure 11.3 shows the block diagram of the project.

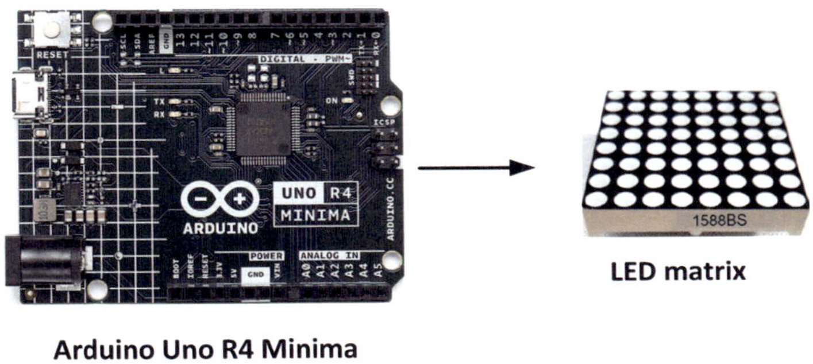

Figure 11.3: Block diagram of the project.

Circuit diagram: The circuit diagram of the project is shown in Figure 11.4, where 8 digital ports are connected to the LED matrix through a 1-kΩ current-limiting resistor. Additionally, 3 pins are used to control the shift register chip.

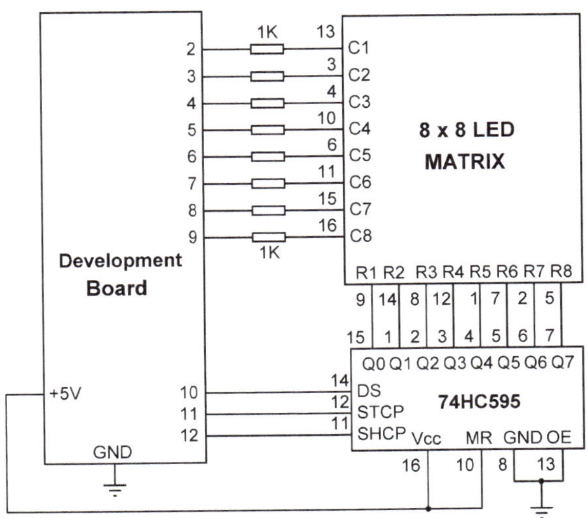

Figure 11.4: Circuit diagram of the project.

Creating the UP ARROW shape:
Creating a shape for the LED matrix is easy. All you have to do is create a shape with 8×8 empty circles. Then fill in the circles to create your shape. Write down the hexadecimal code for the circles that are filled in. Set 0x0 for the empty circle. Notice that each row is represented with a byte. Examples are shown in Figure 11.5.

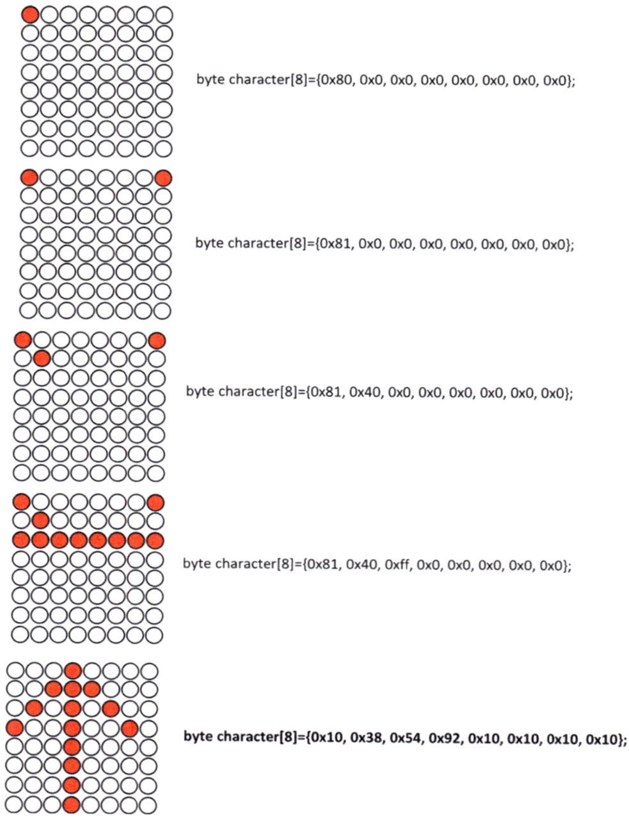

Figure 11.5: Creating shapes for the LED matrix.

The code to generate your up arrow is (last shape in the figure):

```
byte character[8]={0x10, 0x38, 0x54, 0x92, 0x10, 0x10, 0x10, 0x10};
```

Now let's write the program.

Program listing: Figure 11.6 shows the program listing (Program: **LEDMATRIX**). At the beginning of the program, the 74HC595 IC pins are defined, column port pins are defined in array **ColPins**, and bits of the shape to be drawn are stored in array shape. Inside the **setup()** function, 74HC595 pins are configured as outputs. Also, all the column ports are configured as outputs. Function **SendTo595()** receives **data** as its argument and sends it out in parallel form as the display rows.

Inside the main program loop, you select a row and then activate or deactivate the 8-column LEDs belonging to the selected row depending on the shape to be drawn. This process is repeated until all the rows are selected. The display is operated in common anode mode where all te anodes are made HIGH in a row and the LED to be turned ON is made LOW.

```
//---------------------------------------------------------------------------
//                    LED MATRIX
//                    ==========
//
// This program draws an UP ARROW shape on the 8x8 LED matrix
//
// Author: Dogan Ibrahim
// File  : LEDMATRIX
// Date  : June, 2023
//---------------------------------------------------------------------------
int STCP = 11;                                          // Latch
int SHCP = 12;                                          // Clock
int DS = 10;                                            // Data
int ColPins [8] = {2, 3, 4, 5, 6, 7, 8, 9};            // Column pins
byte shape[8]={0x10, 0x38, 0x54, 0x92, 0x10, 0x10, 0x10, 0x10};
int row, col, selrow, i;

void setup()
{
  pinMode(STCP, OUTPUT);                                // Latch is output
  pinMode(SHCP, OUTPUT);                                // Clock is output
  pinMode(DS, OUTPUT);                                  // DS is output

  for(i = 0; i < 8; i++)
  {
    pinMode(ColPins[i],OUTPUT);                         // Column pins output
  }
}

//
// This function writes data to the 74HC595 shift register
//
void SendTo595(byte data)
{
  digitalWrite(STCP, LOW);                              // LAtch LOW
  shiftOut(DS, SHCP, LSBFIRST, data);                   // Shift out
  digitalWrite(STCP, HIGH);                             // LAtch HIGH
}

void loop()
{
```

```
  selrow = 0x80;                                    // Select row 8
  for(row = 0; row < 8; row++)                      // Do for all rows
  {
    for(i = 0; i < 8; i++)
      digitalWrite(ColPins[i], HIGH);               // All cols disabled

    SendTo595(selrow);                              // Enable row

    for(col = 0; col < 8; col++)                    // Do for all cols
    {
      if(shape[row] & 1 << col)                     // Bit set?
        digitalWrite(ColPins[7-col], LOW);          // Enable LED
      else
        digitalWrite(ColPins[7-col], HIGH);         // Disable LED
    }

    selrow = selrow >> 1;                           // Select next row (7,6,..)
    delay(1);                                       // Small delay
  }
}
```

Figure 11.6: Program: ***LEDMATRIX****.*

Figure 11.7 shows the Up arrow displayed on the LED matrix.

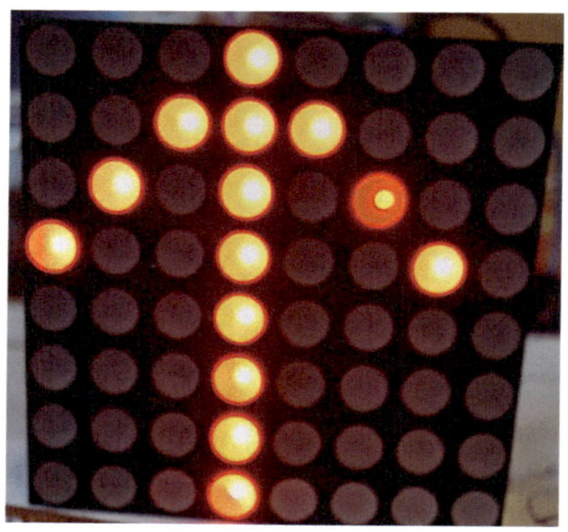

Figure 11.7: Displaying the up arrow.

Chapter 12 • Motors: Servo and Stepper

12.1 Overview

Devices like DC and AC electric motors, stepper motors, and servo motors are actuators typically used to move or rotate objects. Servo motors are frequently used in many hobby and professional electronics applications such as controlling model boats, cars, drones, and aeroplanes remotely. Motors are important in many microcontroller-based projects. For example, in practically all robotic applications, you have to use motors in one way or another. Two types of motors are supplied with the kit: servo motor and stepper motor. In this chapter, you work on projects using both types of motor.

12.2 The servo motor

The servo motor supplied with the kit is the small SG90-type motor manufactured by TowerPro (Figure 12.1).

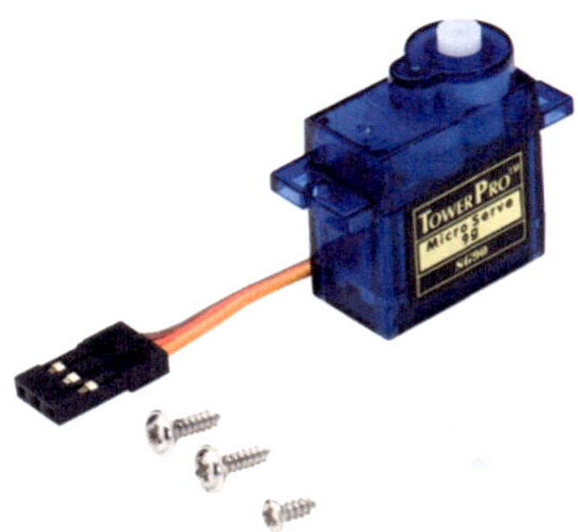

Figure 12.1: Supplied servo motor.

The motor is shipped with a 25-cm length of wire and a 3-pin connector, plus two small plastic propellers. The SG90 servo motor has the following basic specifications:

- Weight: 9 g
- Dimension: 23 × 12.2 × 29 mm
- Stall torque: 1.8 kg/cm
- Gear set: nylon
- Speed: 0.1 sec/60 degrees
- Wires: 3

The motor wiring details are:

Red:	positive supply
Brown:	ground supply
Orange:	control signal

The SG90 servo motor is controlled using PWM waveforms. The SG90 uses high current and because of this, if you are using more than one servo motor with the Arduino, it is important to connect their power connections to an external power supply, as the Arduino may not be able to provide the required current to both motors.

12.2.1 Project 1: Test-rotate the servo

Description: The aim of this project is to learn how to use the servo in a program. In this project, you will perform the following continuously:

- Rotate the servo to 0 degrees.
- Wait 5 seconds.
- Move to 90 degrees.
- Wait 5 seconds.
- Move to 180 degrees.
- Wait 10 seconds.

Block diagram: Figure 12.2 shows the block diagram of the project.

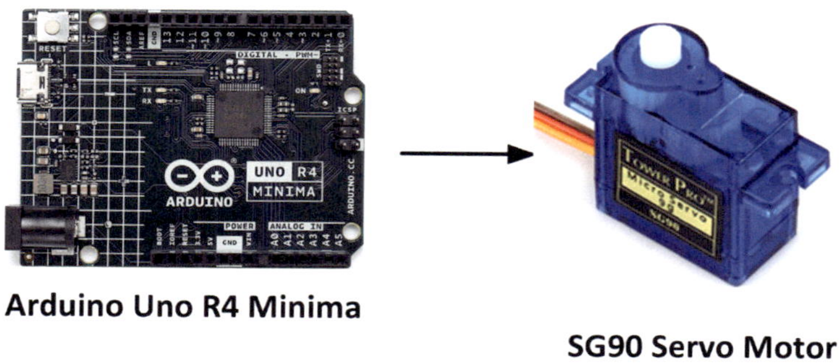

Figure 12.2: Block diagram of the project.

Circuit diagram: Figure 12.3 shows the connections between the servo and the development board ports. Because the servo operates with a PWM waveform, you have to connect it to one of the Arduino Uno PWM ports (e.g., 3, 5, 6, 9, 10, 11). The PWM pins on the Arduino Uno R4 are labelled with the ~ sign. In this project, the selected port is 9.

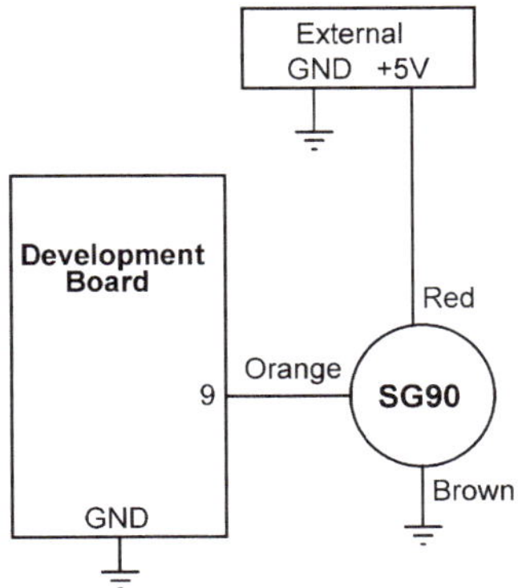

Figure 12.3: Circuit diagram of the project.

Program listing: The projects in this chapter use the servo library. Install the servo library as follows:

- Click to open the **Library Manager** at the left-hand side of the screen.
- Type **Servo**.
- Click to **INSTALL** to install **Servo by Michael Margolis, Arduino**. At the time of drafting this book, the latest version was 1.2.1.

The following functions are commonly used in this library:

attach(pin): attach the servo to a PWM pin
write(angle): set the angle of the servo shaft in degrees
read(): read the current angle of the servo
attached(): check if the servo is attached to a pin
detach(): detach the servo from the specified pin

Figure 12.4 shows the program listing (Program: **Servo1**). At the beginning of the program, the servo library is included and is then attached to port 9 in the **setup()** function. Inside the main program loop, the servo is controlled as required.

```
//----------------------------------------------------------------------
//                      SERVO TEST
//                      ==========
//
// This program controls the servo as described in the text
//
// Author: Dogan Ibrahim
// File   : Servo1
// Date   : June, 2023
//----------------------------------------------------------------------
```

```
#include <Servo.h>
Servo MyServo;

void setup()
{
  MyServo.attach(9);                         // Attach servo to pin 9
}


void loop()
{
  MyServo.write(0);                          // Go to 0 degrees
  delay(5000);                               // Wait 5 seconds
  MyServo.write(90);                         // Go to 90 degrees
  delay(5000);                               // Wait 5 seconds
  MyServo.write(180);                        // Go to 180 degrees
  delay(10000);                              // Wait 10 seconds
}
```

Figure 12.4: Program: **Servo1**.

Note: It is recommended to use an external +5 V supply to power any servo. The current consumption of some SG90 servos can reach 500 mA, exceeding the Arduino's current limit. You can easily measure the current requirements of your servo under different conditions using a multimeter and an external power supply.

12.2.2 Project 2: Servo sweep

Description: This program sweeps the servo lever from 0 to 180 degrees in steps of 10 degrees, with a 5-second delay between each output. The servo lever is then swept back from 180 degrees to 0 degrees after a 3-second delay.

The block diagram and circuit diagram of the project are the same as in Figure 12.2 and Figure 12.3, respectively.

Program listing: Figure 12.5 shows the program listing (Program: **Servo2**). Two **for** loops are used to control the servo shaft. The first loop runs from 0 to 180, while the second one runs from 180 to 0.

```
//----------------------------------------------------------------------
//                    SWEEPING SERVO
//                    ==============
//
// This program sweeps the servo from 0 to 180 degrees in one degrees steps
// and then sweeps back
//
```

```
// Author: Dogan Ibrahim
// File  : Servo2
// Date  : June, 2023
//----------------------------------------------------------------------
#include <Servo.h>
Servo MyServo;
int angle;

void setup()
{
  MyServo.attach(9);                        // Attach servo to pin 9
}


void loop()
{
  for(angle = 0; angle <= 180; angle++)     // Sweep 0 to 180
  {
    MyServo.write(angle);
    delay(100);
  }

  delay(2000);

  for(angle = 180; angle >= 0; angle--)      // Sweep back
  {
    MyServo.write(angle);
    delay(100);
  }
}
```

Figure 12.5: Program: **Servo2**.

12.2.3 Project 3: Joystick-controlled servo

Description: In this program, you control the shaft of the servo using the joystick. Only the horizon movement of the joystick is used. Moving the joystick to the right rotates the motor shaft to the right by 90 degrees, and similarly moving it to the left rotates the shaft to the left by 90 degrees.

Block diagram: Figure 12.6 shows the block diagram of the project.

Figure 12.6: Block diagram of the project.

Circuit diagram: The circuit diagram of the project is shown in Figure 12.7. The joystick is connected as in Chapter 10. The servo is connected to port 9 as in the previous projects. **Although the servo is shown to be connected to +5 V line of the development board, it is recommended to use an external +5 V power supply as in Figure 12.3.**

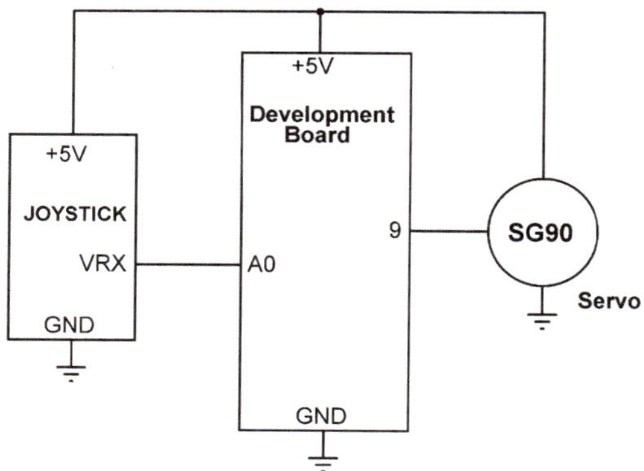

Figure 12.7: Circuit diagram of the project.

Program listing: Figure 12.8 shows the program listing (Program: **Servo3**). The map function is used to map the joystick movements to shaft angles. Notice that the default ADC resolution of 10 bits is used in this project.

```
//----------------------------------------------------------------------
//                    SERVO JOYSTICK CONTROL
//                    ======================
//
// In this program horizontal movements of the joystick controls the
// shaft of the servo motor
//
// Author: Dogan Ibrahim
// File  : Servo3
// Date  : June, 2023
//----------------------------------------------------------------------
#include <Servo.h>
```

```
#define X A0
Servo MyServo;

void setup()
{
  MyServo.attach(9);                        // Attach servo to pin 9
}


void loop()
{
  int Xdir = analogRead(X);                 // Read joystick
  int mapinX = map(Xdir,0, 1023, 180, 0);   // Map
  MyServo.write(mapinX);                    // Move servo
  delay(50);
}
```

Figure 12.8: Program: **Servo3**.

Note: by changing the mapping to **int mapinX = map(Xdir, 0, 512, 180, 0)** you can move the shaft 0 to 180 degrees left by moving the joystick from its idle position to fully left.

12.3 The stepper motor

Stepper motors are brushless synchronous motors where a full rotation is divided into many steps. The motor rotates by a single step when a step pulse is applied to it. These motors are manufactured with steps of 12, 24, 36, 72, 144, 180, etc. steps per full revolution. With a 12-step motor, the stepping angle is 30 degrees so that it makes a complete 360 degrees revolution in 12 steps.

The stepper motor supplied with the kit (Figure 12.9) is the type 28BYJ-48. Each step of this motor equals 12.5 degrees, and therefore it takes 32 steps to complete a full revolution. Additionally, this motor has a 1/64 gear set, which means that there are actually 32 × 64 = 2038 steps per revolution. The current requirement of the motor is about 240 mA which vastly exceeds the current capacity of an I/O pin of Arduino Uno. It is therefore **mandatory** to power the motor from an external 5-V power supply.

The 28BYJ-48 motor is supplied with the ULN2003 drive board (see Figure 12.9). You should connect the motor to this driver module before using it. The driver board has 4 on-board LEDs which indicate the steps applied to the motor. Four control inputs labelled IN1 to IN4 are provided at the edge of the board for connection to the processor. An ON/OFF jumper is also available on the board to isolate power from the board.

Figure 12.9: 28BYJ-48 stepper motor and driver board.

12.3.1 Project 4: Rotate the motor clockwise and then anticlockwise

Description: The aim of this project is to show how to use the stepper motor in a project. In this project, the stepper motor is rotated slowly clockwise, stopped for 2 seconds, and then rotated anticlockwise continuously.

Block diagram: Figure 12.10 shows the block diagram of the project.

Figure 12.10: Block diagram of the project.

Circuit diagram: The circuit diagram of the project is shown in Figure 12.11. Ports 2, 3, 4, 5 are used to control the motor. An external +5 V power supply must be used to provide power to the board through the driver board. Connect the +5 V line of the power supply to pin marked **+** and the GND to pin marked **–** on the driver board.

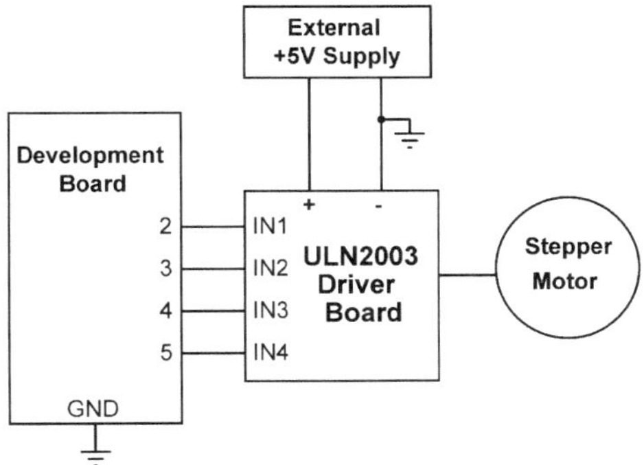

Figure 12.11: Circuit diagram of the project.

Program listing: In this program, the Arduino IDE stepper motor library is used. Install the library as follows:

- Click to open the **Library Manager** at the left-hand side of the screen.
- Type **Stepper**.
- Click to **INSTALL** to install **Stepper by Arduino**. At the time of drafting this book, the latest version was 1.1.3.

The commonly used functions supported by this library are:

stepper(steps, in1, in2, in3, in4): create an instance of the Stepper class. Define the number of steps per revolution and the connections to the motor.

setSpeed(rpms): set the motor speed in rotations per minute (rpms). This function doesn't make the motor turn, just sets the speed at which it will when you call **step().**

step(steps): this function turns the motor a specific number of steps, at a speed determined by the most recent call to **setSpeed()**. This function is blocking; that is, it will wait until the motor has finished moving to pass control to the next line in your program.

The maximum speed for the 28BYJ-48 stepper motor is about 10-15 rpm when operated at +5 V.

Figure 12.12 shows the program listing (Program: **STEPPER1**). At the beginning of the program, the stepper library is included, steps per revolution is defined and the **Stepper** instance is created. Notice that the driver board wirings must be declared in the software in the order of: **IN1**, **IN3**, **IN2** and **IN4**. Inside the **setup()** function, the speed of the motor is set to 10 RPM. Inside the main program loop, the motor rotates one full revolution clockwise, waits 2 seconds, and then rotates one full revolution anticlockwise. This is repeated forever after 2 seconds of delay. Notice that the **step()** function is blocking and it will wait until the motor has finished moving.

```
//--------------------------------------------------------------------
//                    STEPPER MOTOR CONTROL
//                    =====================
//
// In this program the stepper motor rotates cw direction and then
// after a short delay ccw direction. The maximum speed of the supplied
// motor is about 10-15 rpm
//
// Author: Dogan Ibrahim
// File  : STEPPER1
// Date  : June, 2023
//--------------------------------------------------------------------
#include <Stepper.h>

int StpsPerRev = 2038;                                // Steps per rev
Stepper MyStepper = Stepper(StpsPerRev,2,4,3,5);     // IN1,3,2,4

void setup()
{
  MyStepper.setSpeed(10);                             // 10 RPM
}


void loop()
{
  MyStepper.step(StpsPerRev);                         // cw direction
  delay(2000);

  MyStepper.step(-StpsPerRev);                        // CCW direction
  delay(2000);
}
```

Figure 12.12: Program: **STEPPER1**.

Chapter 13 • The Digital To Analog Converter (DAC)

13.1 Overview

A DAC (digital-to-analog converter) converter is useful for converting digital signals into analog form. The Arduino Uno R4 supports one DAC. For some applications where an analog output is required, you can use the PWM (Pulse Width Modulation) instead of "genuine" analog output. For example, when dimming an LED, you can simply use a PWM-enabled digital pin as an analog output pin and the LED would dim just the same as if you'd be using a DAC output. For many audio-based applications as well as for wave generation, it is required to use genuine analog output. In this chapter, however, DAC-based projects are discussed.

The default resolution of the DAC is 8 bits (0–255), but it can be changed to 12 bits (0–4096) with the following statement. The analog output is available at pin A0 (or DAC):

```
analogWriteResolution(12);
```

13.2 Project 1: Generating a square wave with 2 V amplitude

Description: In this project, a square wave is generated at the DAC output with a frequency of 500 Hz (Period: 2 ms) and an amplitude of 2 V. The aim of this project is to show how the DAC can be used in a simple application.

Program listing: Figure 13.1 shows the program listing (Program: **Square2V**). The DAC is operated at default 8 bits. With a reference voltage of 5 V, 2 V corresponds to 2 × 256 / 5 = 102. In this program, the **millis()** function is used to create an accurate delay instead of the **delay()** function. The program repeatedly sends 102 to DAC which corresponds to 2 V, waits for 1 ms, and then sends 0. The output waveform is plotted on an OWON digital oscilloscope as shown in Figure 13.2 (connect the oscilloscope to pin A0 of the development board). In this figure, the horizontal axis was 500 microseconds/division, and the vertical axis was 2 V/division. The waveform has an amplitude of 2 V and a period of 2 ms.

```
//----------------------------------------------------------------------------
//                 SQUARE WAVE WITH AMPLITUDE 2 V
//                 ==============================
//
// This program generates square wave with frequency 500 Hz and amplitude 2 V
//
// Author: Dogan Ibrahim
// File  : Square2V
// Date  : July, 2023
//----------------------------------------------------------------------------
unsigned long previousMillis = 0;
const long interval = 1;
bool flag = false;

void setup()
```

```
{
}

void loop()
{
  unsigned long currentMillis = millis();
  if(currentMillis - previousMillis >= interval)
  {
    previousMillis = currentMillis;
    if(flag)
    {
      analogWrite(A0, 102);
      flag = false;
    }
    else
    {
      analogWrite(A0, 0);
      flag = true;
    }
  }
}
```

Figure 13.1: Program: **Square2V**.

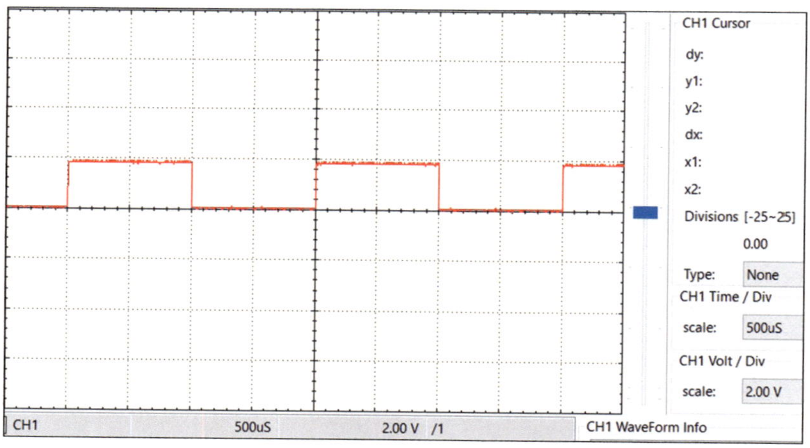

Figure 13.2: Output waveform.

13.3 Generating sine wave – using the analogWave library

The analogWave library contains several functions and is used to generate various waveforms through the DAC. The waveform is being stored as samples in an array, and with every loop of the sketch, you'll update the DACs output value to the next value in the array. With predefined waveforms, the library is called as: **analogWave wave(DAC)**. The following functions are available:

- begin(float freq_hz) e.g. wave.begin(100) - commence the output
- freq(float freq_hz) - update the frequency
- start() - start the generation of sample
- stop() - stop the generation of sample
- amplitude(float amplitude) - output multiplication, between 0 and 1
- sine(float freq_hz) e.g. wave.sine(100) - sine wave
- square(float freq_hz) - square wave
- saw(float freq_hz) - sawtooth wave

Example projects are given in the next sections.

13.3.1 Project 2: Generate a sine wave

Description: In this project, you and your R4 generate a sine wave with a frequency of 1 kHz and an amplitude of 2.5 V.

Program listing: Figure 13.3 shows the program listing (Program: **Sinewave**). At the beginning of the program, the frequency is set to 1 kHz (1000 Hz). The amplitude multiplier is set to 0.5 so that the peak-to-peak amplitude of the generated waveform will be 5 × 0.5 = 2.5 V (notice that the DAC reference voltage is not exactly +5 V, it is slightly lower). Inside the **setup()** function, the amplitude is set and waveform is generated. Figure 13.4 shows the waveform on the digital oscilloscope. In this figure, the vertical scale was 1 V/division, and the horizontal scale was 500 µs/division. The frequency is 1 kHz, and the amplitude is about 2.5 V.

```
//----------------------------------------------------------------------------
//                  GENERATE SINE WAVE
//                  ==================
//
// This program generates sine wave with frequency 1 kHz and amplitude 2.5 V
//
// Author: Dogan Ibrahim
// File  : Sinewave
// Date  : July, 2023
//----------------------------------------------------------------------------
#include "analogWave.h"
analogWave wave(DAC);

int freq = 1000;                                    // Freq = 1 kHz
float amplitude = 0.5;                              // Amplitude multiplier

void setup()
{
  wave.amplitude(amplitude);                        // Set Amplitude
  wave.sine(freq);                                  // Generate sine wave
}
```

```
void loop()
{
}
```

Figure 13.3: Program: ***Sinewave.***

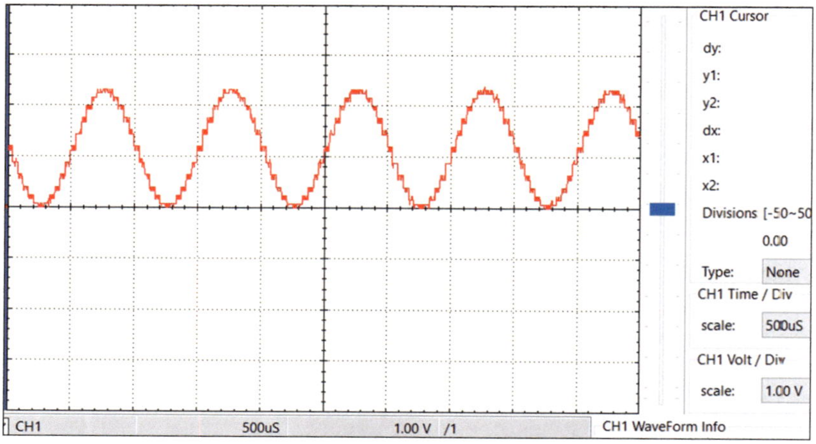

Figure 13.4: Output waveform.

Notice that in the above program, the default DAC resolution of 8 bits was used. The resolution can be changed to 12 bits to improve the waveform:

```
void setup()
{
        analogWriteResolution(12);                  // Change to 12-bits
        wave.amplitude(amplitude);                  // Set Amplitude
        wave.sine(freq);                            // Generate sine wave
}
```

13.3.2 Project 3: Sine wave sweep frequency generator

Description: In this project, a sine wave is generated with the frequency changing from 100 Hz to 10 kHz in steps of 100 Hz, every second. The amplitude of the waveform is 5 V by default.

Program listing: Figure 13.5 shows the program listing (Program: **Sweep**). Here, the start frequency is 100 Hz and the frequency is increased in steps of 100 Hz every second inside the **loop()** function. When the frequency reaches 10 kHz, it stays there for 5 seconds and then the process repeats.

```
//----------------------------------------------------------------------------
//                SINE WAVE SWEEP FREQUENCY
//                =========================
//
// This program generates sine wave starting from 100 Hz to 10 kHz in steps
// of 100 Hz. The frequncy is incremented every second. The wave amplitude
// is default 5 V
//
// Author: Dogan Ibrahim
// File  : Sweep
// Date  : July, 2023
//-----------------------------------------------------------------------------
#include "analogWave.h"
analogWave wave(DAC);

int StartFreq = 100;                                   // Start freq = 100 Hz
int EndFreq = 10000;                                   // End freq = 10 kHz
int Step = 100;                                        // Step = 100 Hz
int freq = StartFreq;

void setup()
{
  wave.sine(StartFreq);                                // Start with StartFreq
}

void loop()
{
  freq = freq + Step;                                  // Increment freq
  delay(1000);

  if(freq > EndFreq)                                   // Reached end?
  {
    freq = StartFreq;                                  // Back to start
    delay(5000);
  }
  wave.freq(freq);                                     // With new freq
}
```

Figure 13.5: Program: ***Sweep****.*

13.3.3 Project 4: Generate sine wave whose frequency changes with potentiometer

Description: In this program, a potentiometer is connected to the Uno R4's analog pin A5. A sine wave is generated, and its frequency changes as the potentiometer arm is moved.

Circuit diagram: Figure 13.6 shows the circuit diagram of the project. The arm of the potentiometer is connected to the analog input A5 of the development board.

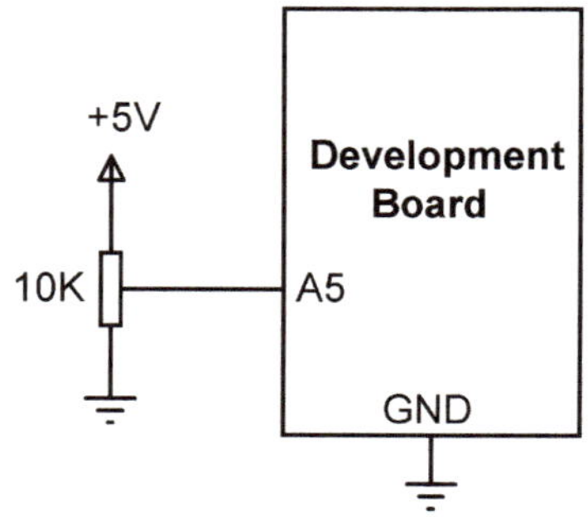

Figure 13.6: Circuit diagram of the project.

Program listing: Figure 13.7 shows the program listing (Program: **DACPot)**. The ADC and DAC are left with their default resolutions. Inside the program loop, the value of the potentiometer is read and mapped to 0 to 10000 Hz. The current frequency is then generated at port A0 (DAC) and is also displayed on the Serial Monitor.

```
//---------------------------------------------------------------------------
//                SINE WAVE WITH A POTENTIOMETER
//                ==============================
//
// This program generates sine wave where a potentiometer is used to change
// the frequency. The wave amplitude is default 5 V
//
// Author: Dogan Ibrahim
// File  : DACPot
// Date  : July, 2023
//---------------------------------------------------------------------------
#include "analogWave.h"
analogWave wave(DAC);

int freq = 100;                                      // Start freq = 100 Hz

void setup()
{
  Serial.begin(9600);
  delay(5000);
  pinMode(A5, INPUT);                                // Analog input
  wave.sine(freq);                                   // Start with StartFreq
}

void loop()
{
```

```
  freq = map(analogRead(A5), 0, 1024, 0, 10000);
  Serial.println("Frequency is: " + String(freq) + " Hz");
  wave.freq(freq);
  delay(1000);
}
```

Figure 13.7: Program: **DACPot**.

13.3.4 Project 5: Generate a square wave with frequency of 1 kHz and amplitude of 1 V

Description: In this project, a 1-kHz square wave is generated with an amplitude of 1 V.

Program listing: Figure 13.8 shows the program listing (Program: **Square1V)**. The amplitude is set to 1 V by using a 0.2 output level multiplier. The default DAC resolution of 8 bits is retained in this project.

```
//---------------------------------------------------------------------------
//                GENERATE SQUARE WAVE
//                ====================
//
// This program generates square wave with frequency 1 kHz and amplitude 1 V
//
// Author: Dogan Ibrahim
// File  : Square1V
// Date  : July, 2023
//----------------------------------------------------------------------------
#include "analogWave.h"
analogWave wave(DAC);

int freq = 1000;                                    // Freq = 1 kHz
float amplitude = 0.2;                              // Amplitude multiplier

void setup()
{
  wave.amplitude(amplitude);                        // Set Amplitude
  wave.square(freq);                                // Generate square wave
}

void loop()
{
}
```

Figure 13.8: Program: **Square1V**.

Suggestion: A simple decent quality melody maker can be constructed by using the DAC together with an audio amplifier like the LM386 IC and a small 4–8 ohm speaker.

Chapter 14 • Using the EEPROM, the Human Interface Device, and PWM

14.1 Overview

Both versions of the Arduino Uno R4 support some additional features, such as EEPROM memory and a Human Interface Device (HID). In this chapter, these features are described with simple examples.

14.2 The EEPROM memory

The EEPROM is useful when it is required to store non-volatile data such as passwords, initialization data, etc. Some of the commonly used EEPROM library functions are as follows:

```
EEPROM.read(int address)
```

This function allows you to read a single byte of data from the EEPROM. An **unsigned char** is returned.

```
EEPROM.write( int address, unsigned char value )
```

This function allows you to write a single byte of data to the EEPROM. The first parameter is the address to be written, and the second parameter is the data to be written. The function does not return any data.

```
EEPROM.length()
```

This function returns an **unsigned int** containing the number of cells in the EEPROM.

The following program code will write byte 0x11 and byte 0x22 to EEPROM addresses 0 and 1 respectively:

```
#include <EEPROM.h>
int addr = 0;
EEPROM.write(addr, 0x11);
addr++;
EEPROM.write(addr, 0x22);
```

The following program code will read the byte at EEPROM address 0 and then display it on the Serial Monitor. The number is displayed in decimal together with its address:

```
#include <EEPROM.h>
Serial.begin(9600);
int addr = 0;
byte value;
value = EEPROM.read(addr);
Serial.print(addr);
Serial.print("\t");
```

```
Serial.print(value, DEC);
Serial.println();
```

14.3 Human Interface Device (HID)

The HID can be used to emulate the mouse/keyboard using the Arduino Uno R4 Minima board in conjunction with the keyboard and mouse APIs. HIDs are devices designed for humans (keyboards, mice, game controllers, etc.), that frequently send data to a computer over USB. When you press a key on a keyboard, you send data to a computer, which reads it and in turn activates the corresponding key. To turn your board into an HID, you can use the keyboard/mouse API built into the core. This chapter is an introduction to HID and further information and examples on HID can be obtained from the following Arduino tutorial:

https://docs.arduino.cc/tutorials/uno-r4-minima/usb-hid

Some of the keyboard functions include **begin()**, **press()** (pressing a key), and **releaseAll()** (releasing keys). An example program code is given below for the keyboard. This code emulates a keypress and key release. The following code prints the letter **x** every second:

```
#include <Keyboard.h>
void setup()
{
        Keyboard.begin()
        delay(1000);
}

void loop()
{
        Keyboard.press('x');
        delay(100);
        Keyboard.releaseAll();
        delay(1000);
}
```

Some of the mouse functions are: **begin()**, and **move(xDistance, yDistance), press(MOUSE_LEFT), release(MOUSE_LEFT)**. As an example, the following example moves both axes of the mouse just slightly (20 points), back and forth every second:

```
#include <Mouse.h>
void setup()
{
        Mouse.begin();
delay(1000);
}

void loop()
{
```

```
        Mouse.move(20, 20);
        delay(1000);
        Mouse.move(-20, -20);
        delay(1000);
}
```

An example project is given below using the Keyboard library.

14.4 Project 1: Keyboard control to launch Windows programs

Description: This project uses the Keyboard library. Three buttons are connected to the development board, named **Notepad**, **Word**, and **Excel**. Pressing the Notepad button starts the Notepad program, pressing the Word button starts the Winword program, and pressing the Excel button starts the Excel program. The aim of this project is to show how the Keyboard library can be utilized.

Block diagram: Figure 14.1 shows the block diagram of the project.

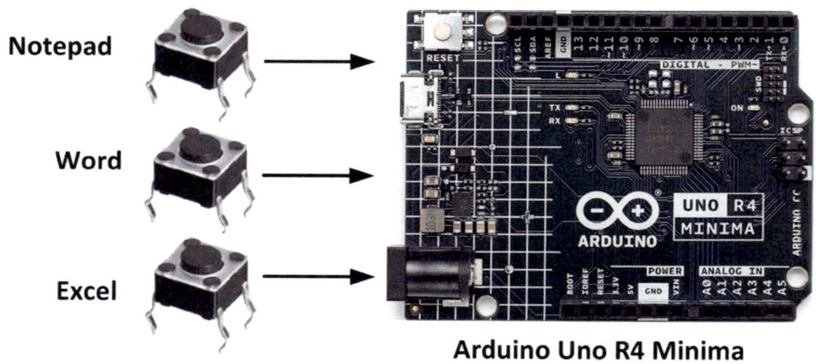

Figure 14.1: Block diagram of the project.

Circuit diagram: The circuit diagram is shown in Figure 14.2 Buttons **Notepad**, **Word**, and **Excel** are connected to digital pins 2, 3, and 4 respectively.

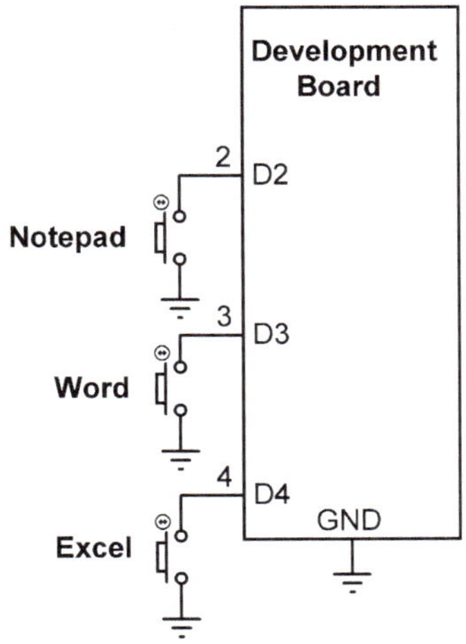

Figure 14.2: Circuit diagram.

Program listing: Figure 14.3 shows the program listing (Program: **keybd**). At the beginning of the program, the Keyboard library header file is included, and buttons are assigned to digital ports. Inside the **setup()** function, the buttons are configured as inputs and the input pull-up resistors are enabled so that the default state of the buttons is at logic HIGH. The state of a button goes to logic LOW when pressed. Inside the main program loop, the state of the buttons is checked. If the button **Notepad** is pressed, then the variable **Mode** is set to 1. If the button **Word** is pressed, then the variable **Mode** is set to 2, and if the button **Excel** is pressed, then the variable **Mode** is set to 3. The program then sends the left **Windows key** to the PC, followed by letter **r** (Run) and then the name of the program to run. For example, to run the **Notepad** program, the keystrokes are **Windows key**, followed by **r** key, followed by the word **Notepad** and then the **Enter** (Return) key. So, for example, pressing the **Excel** button launches Excel to run within Windows.

```
//---------------------------------------------------------------------
//             KEYBOARD CONTROL TO START WINDOWS PROGRAMS
//             ==========================================
//
// Three buttons are used in this program named Notepad, Word, and Excel.
// Pressing button Notepad starts notepad, pressing Word starts Winword,
// and pressing Excel starts the Excel program
//
// Author: Dogan Ibrahim
// File  : Keybd
// Date  : July, 2023
//----------------------------------------------------------------------
#include <Keyboard.h>

#define ButtonNotePad 2                          // Notepad button
#define ButtonWord 3                             // Word button
```

```
#define ButtonExcel 4                              // Excel button
int Mode;

void setup()
{
  pinMode(ButtonNotePad, INPUT_PULLUP);            // Default 1
  pinMode(ButtonWord, INPUT_PULLUP);               // Default 1
  pinMode(ButtonExcel, INPUT_PULLUP);              // Default 1
  Keyboard.begin();                                // Initialize library
}


void loop()
{
  Mode = 0;

  while(Mode == 0)
  {
    if(digitalRead(ButtonNotePad) == LOW) Mode = 1;   // Notepad pressed
    if(digitalRead(ButtonWord) == LOW) Mode = 2;      // Word pressed
    if(digitalRead(ButtonExcel) == LOW) Mode = 3;     // Excel pressed
  }

  Keyboard.press(KEY_LEFT_GUI);                       // Windows key
  Keyboard.press('r');                                // Run
  delay(100);
  Keyboard.releaseAll();
  delay(1000);
  if(digitalRead(Mode) == 1) Keyboard.print("Notepad");
  if(digitalRead(Mode) == 2) Keyboard.print("Winword");
  if(digitalRead(Mode) == 1) Keyboard.print("Excel");
  Keyboard.press(KEY_RETURN);
  delay(100);
  Keyboard.releaseAll();
 }
```

Figure 14.3: Program: ***keybd****.*

Figure 14.4 shows the project built on a breadboard and the buttons connected to the development board using jumper wires.

Figure 14.4: Construction of the project.

Some key names are given below. The full list can be obtained from the website:

https://github.com/arduino-libraries/Keyboard/blob/master/src/Keyboard.h

KEY_LEFT_CTRL	left CTRL key
KEY_LEFT_SHIFT	left SHIFT key
KEY_LEFT_ALT	left ALT key
KEY_LEFT_GUI	left WINDOWS key
KEY_RIGHT_CTRL	right CTRL key
KEY_RIGHT_SHIFT	right SHIFT key
KEY_RIGHT_ALT	right ALT key
KEY_RIGHT_GUI	right WINDOWS key
KEY_UP_ARROW	UP arrow key
KEY_DOWN_ARROW	DOWN arrow key
KEY_LEFT_ARROW	LEFT arrow key
KEY_RIGHT_ARROW	RIGHT arrow key
KEY_RETURN	RETURN (ENTER) key
KEY_ESC	ESC key
KEY_INSERT	INSERT key
KEY_DELETE	DELETE key

14.5 The Pulse Width Modulation (PWM)

Pulse Width Modulation (PWM) is a commonly used technique for controlling the power delivered to analog loads using digital waveforms. Although analog voltages (and currents) can be used to control the delivered power, they have several drawbacks. Controlling large analog loads requires large voltages and currents that cannot easily be obtained using standard analog circuits and DACs. Precision analog circuits can be heavy, large, expensive, and sensitive to noise. By using the PWM technique, the average value of voltage (and current) fed to a load is controlled by switching the supply voltage ON and OFF at a fast rate. The longer the power on time, the higher the voltage supplied to the load.

Figure 14.5 shows a typical PWM waveform where the signal is basically a repetitive positive pulse, having the period T, ON time TON and OFF time of T – TON seconds. The minimum and maximum values of the voltage supplied to the load are 0 and VP respectively. The PWM switching frequency is usually set to be very high (usually in the order of several kHz) so that it does not affect the load that uses the power. The main advantage of PWM is that the load is operated efficiently since the power loss in the switching device is very low. When the switch is ON there is practically no voltage drop across the switch, and when the switch is OFF there is no current supplied to the load.

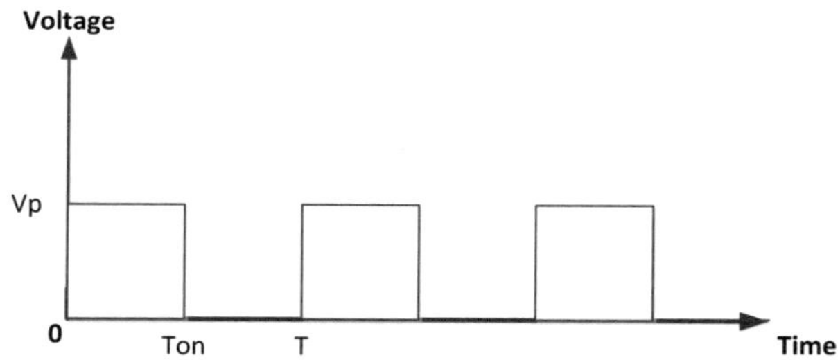

Figure 14.5: PWM waveform.

The duty cycle (or D) of a PWM waveform is defined as the ratio of the ON time to its period. Expressed mathematically,

Duty Cycle (D) = TON / T

The duty cycle is usually expressed as a percentage and therefore,

D = (TON / TOFF) × 100 %

By varying the duty cycle between 0% and 100% you can effectively control the average voltage supplied to the load between 0 and Vp.

The average value of the voltage applied to the load can be calculated by considering a general PWM waveform shown in Figure 14.6. The average value A of waveform f(t) with period T and peak value ymax and minimum value ymin is calculated as:

$$A = \frac{1}{T}\int_{0}^{T} f(t)\,dt$$

or

$$A = \frac{1}{T}\left(\int_{0}^{T_{ON}} y_{\max}\,dt + \int_{T_{ON}}^{T} y_{\min\,dt}\right)$$

In a PWM waveform ymin = 0 and the above equation becomes

$$A=\frac{1}{T}\left(T_{ON}\ y_{max}\right)$$

or $$A=D\ y_{max}$$

As can be seen from the above equation, the average value of the voltage supplied to the load is directly proportional to the duty cycle of the PWM waveform, and by varying the duty cycle you control the average load voltage. Figure 14.6 shows the average voltage for different values of the duty cycle.

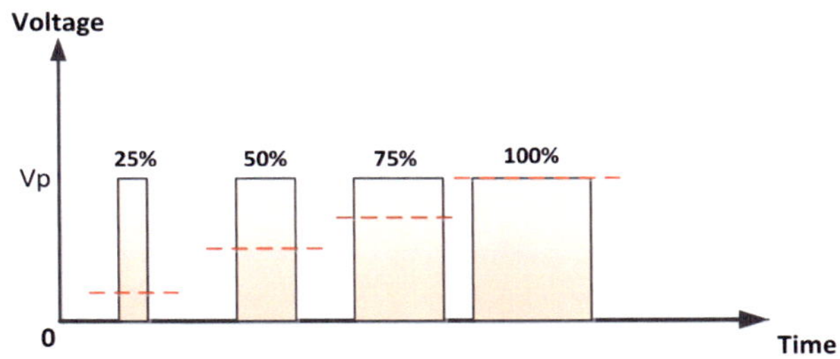

Figure 14.6: Average voltage (shown as dashed line) supplied to a load.

It is interesting to notice that with correct low-pass filtering, the PWM can be used as a DAC if the MCU does not have a DAC channel. By varying the duty cycle, you can effectively vary the average analog voltage supplied to the load.

14.5.1 PWM channels of the Arduino Uno R4

The PWM channels on the Arduino Uno R4 are identified by the ~ (tilde) sign. These channels are: D3, D5, D6, D9, D10, D11. The PWM channels are configured by specifying the required frequency and duty cycle. The duty cycle can be set between 0% and 100% by setting it from 0 to 255. An example project is given in the next section to show how the PWM can be programmed.

14.5.2 Project 2: LED dimming using PWM

Description: In this project, an external LED is connected to port D3 through a 1-kΩ current-limiting resistor. The LED brightness is changed using PWM.

Program listing: Figure 14.7 shows the program listing (Program: **LEDDim**). At the beginning of the program, the brightness (variable **brightness**) and amount of fading (variable **amount**) are both set to 0 and the LED is configured as an output. Inside the program loop, a PWM waveform is sent to the LED with the duty cycle set to **brightness**, and the brightness is increased by the **amount**. When **brightness** reaches the value 255 (maximum duty cycle), the **amount** is subtracted from the **brightness**. This process continues after a 50-ms delay and until stopped by the user. Try to change the delay to see the change in the speed of fading.

```
//--------------------------------------------------------------------------
//                LED DIM - USING PWM
//                ===================
//
// In this program an LED is connected to port 3 and it is dimmed using PWM
//
// Author: Dogan Ibrahim
// File  : LEDDim
// Date  : July, 2023
//---------------------------------------------------------------------------
#define LED 3                                  // LED at port 3
int brightness = 0;
int amount = 10;

void setup()
{
  pinMode(LED, OUTPUT);                        // LED is output
}

void loop()
{
  analogWrite(LED, brightness);
  brightness = brightness + amount;
  if(brightness <= 0 || brightness >= 255) amount = -amount;
  delay(50);
}
```

Figure 14.7: Program: **LEDDim**.

Chapter 15 • The Arduino Uno R4 WiFi

15.1 Overview

The Arduino Uno R4 WiFi (Figure 15.1) is the WiFi version of the Uno R4. It offers some features not found on the Arduino Uno R4 Minima.

Compared to the Arduino Uno R4 Minima , the features only available on the **Arduino Uno R4 WiFi** can be summarized as

- WiFi 2.4 GHz, 802.11b/g/n, up to 150 Mbits/s
- Bluetooth LE and Bluetooth 5
- Built-in antenna
- 12×8 LED matrix
- Connector to add an external battery to the real-time clock
- Qwicc-style connector for 3.3-V level I^2C (additional channel from the MCU)
- Secondary 240 MHz, 3.3-V dual-core 32-bit Espressif ESP32-S3-MINI-1-N8 processor
- 384 Kbytes ROM
- 512 Kbytes RAM
- On-board programmable header
- VRTC pin for external RTC battery

Figure 15.1: Arduino Uno R4 WiFi.

Although the WiFi and Minima share the same pinout, some of the special pin functions are different (e.g., the CAN bus pins). The Uno R4 Minima is programmed directly over USB, while on the WiFi version the MCU is programmed via the ESP32 processor (the USB bus is connected directly to ESP32).

Also, on the Uno R4 Minima the MCU debug is brought out to the SWD connector to connect to an external debugger. On the Uno R4 WiFi, the MCU debug port is taken to the ESP32 processor. The Uno R4 WiFi includes an error-checking mechanism that can detect run-time errors and crashes, and it can provide detailed explanations about the line of code causing the crash.

By default, the ESP32-S3 module on-board the UNO R4 WiFi acts as a Serial Bridge, handling the connection to your computer. It also handles the rebooting of the main MCU, the Renesas RA4M1 when needed, for example, when receiving a new sketch and resetting. The ESP32 also exposes the ESP32's data lines, enabling you to program the ESP32 directly. These data lines are exposed by 3×2 header at the top of the board, or through pads at the bottom side. The USB data lines are routed through switches. By default, these switches are set for communication to go via the ESP32 module. This, however, can be modified by pulling pin D40 HIGH using the **digitalWrite()** statement. While D40 is HIGH, the RA4M1 is connected to the USB Serial port, and while D40 is LOW, the ESP32 is connected, like the default configuration.

The built-in antenna is shared between WiFi and Bluetooth, meaning you cannot use WiFi and Bluetooth at the same time.

Further details on the Arduino Uno R4 WiFi are available at the following Arduino link:

https://docs.arduino.cc/tutorials/uno-r4-wifi/cheat-sheet

Figure 15.2 shows the pin layout of the Arduino Uno R4 WiFi. The component layout on the board is shown in Figure 15.3 with Table 15.1 describing the components on the board. Figure 15.2, Figure 15.3, and Table 5.1 are taken from the ***Arduino Product Reference Manual: SKU: ABX00087***.

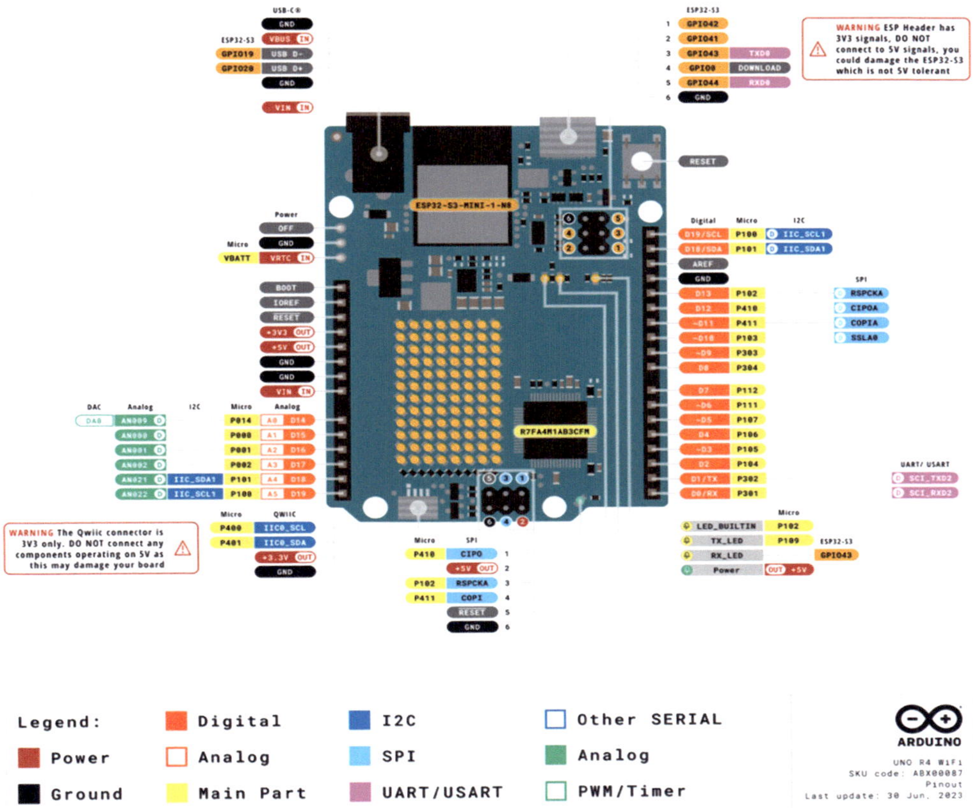

Figure 15.2: Arduino Uno R4 WiFi pin layout.

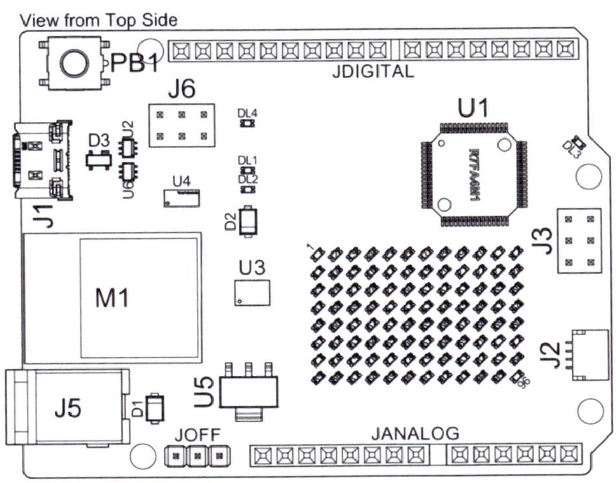

Figure 15.3: Arduino Uno R4 WiFi component layout.

Ref.	Description
U1	R7FA4M1AB3CFM#AA0 Microcontroller IC
U2	NLASB3157DFT2G Multiplexer
U3	ISL854102FRZ-T Buck Converter
U4	TXB0108DQSR logic level translator (5 V - 3.3 V)
U5	SGM2205-3.3XKC3G/TR 3.3 V linear regulator
U6	NLASB3157DFT2G Multiplexer
U_LEDMATRIX	12x8 LED Red Matrix
M1	ESP32-S3-MINI-1-N8
PB1	RESET Button
JANALOG	Analog input/output headers
JDIGITAL	Digital input/output headers
JOFF	OFF, VRTC header
J1	CX90B-16P USB-C® connector
J2	SM04B-SRSS-TB(LF)(SN) I2C connector
J3	ICSP header (SPI)
J5	DC Jack
J6	ESP header
DL1	LED TX (serial transmit)

Table 15.1: Description of components on the board.

Example projects are given in this chapter on using some of the features of the Uno R4 WiFi development board.

15.2 The LED matrix

The LED matrix is organized as 12×8 LEDs and can be used to display graphics, animations, or play games. Two methods are suggested for programming the LED matrix. Both methods are described with projects in this section.

15.2.1 Project 1: Using LED matrix 1 — creating a large + shape

Description: In this project, you will create a large **+** shape on the LED matrix. The aim of this project is to show how a shape can be created on the LED matrix.

Program listing: The LED Matrix library works on the principle of creating a frame (an image), and then loading it into a buffer that displays the frame. In order to control the LED matrix, you need a space in memory sized at least 96 bits.

The required image is created as a two-dimensional array of bytes. For the **+** shape, the array looks as in Figure 15.4:

```
byte frame[8][12] = {
  {0, 0, 0, 0, 0, 1, 0, 0, 0, 0, 0, 0 },
  {0, 0, 0, 0, 0, 1, 0, 0, 0, 0, 0, 0 },
  {0, 0, 0, 0, 0, 1, 0, 0, 0, 0, 0, 0 },
  {1, 1, 1, 1, 1, 1, 1, 1, 1, 1, 1, 1 },
  {0, 0, 0, 0, 0, 1, 0, 0, 0, 0, 0, 0 },
  {0, 0, 0, 0, 0, 1, 0, 0, 0, 0, 0, 0 },
  {0, 0, 0, 0, 0, 1, 0, 0, 0, 0, 0, 0 },
  {0, 0, 0, 0, 0, 1, 0, 0, 0, 0, 0, 0 }
};
```

Figure 15.4: The array of bytes.

The image can easily be seen in the frame of the array as the ones form the required image. You will then have to call function **renderBitmap()**. Figure 15.5 shows the program listing (Program: **MatrixPlus**). At the beginning of the program, the LED matrix header file is included and a matrix object is created. Inside the **setup()** function, the LED matrix is started and the frame is defined and rendered.

```
//----------------------------------------------------------------------------
//                  USING THE LED MATRIX - DISPLAYING + SIGN
//                  ========================================
//
// This program displays a + sign on the LED matrix
//
// Author: Dogan Ibrahim
// File  : MatrixPlus
// Date  : July, 2023
//-----------------------------------------------------------------------------
#include "Arduino_LED_Matrix.h"
ArduinoLEDMatrix matrix;

void setup()
{
  matrix.begin();

  byte frame[8][12] =
  {
  { 0, 0, 0, 0, 0, 1, 0, 0, 0, 0, 0, 0 },
  { 0, 0, 0, 0, 0, 1, 0, 0, 0, 0, 0, 0 },
  { 0, 0, 0, 0, 0, 1, 0, 0, 0, 0, 0, 0 },
  { 1, 1, 1, 1, 1, 1, 1, 1, 1, 1, 1, 1 },
  { 0, 0, 0, 0, 0, 1, 0, 0, 0, 0, 0, 0 },
  { 0, 0, 0, 0, 0, 1, 0, 0, 0, 0, 0, 0 },
  { 0, 0, 0, 0, 0, 1, 0, 0, 0, 0, 0, 0 },
  { 0, 0, 0, 0, 0, 1, 0, 0, 0, 0, 0, 0 }
  };

  matrix.renderBitmap(frame, 8, 12);
```

```
}

void loop()
{
}
```

Figure 15.5: Program: **MatrixPlus**.

Figure 15.6 shows the display.

Figure 15.6: The display.

Notice that you could set all the bits in the frame to 0 and then set the required bits to 1 using the following statements (the array indexing starts from 0):

frame[1, 5] = 1; and so on for all the bits to be set to 1s.

....

....

matrix.renderBitmap(frame, 8, 12);

An example project is given below which sets the required bits in the frame.

15.2.2 Project 2: Creating images by setting bits

Description: In this project, two simple images are created by setting the bits in frames having all 0s.

Program listing: Two images created in this project: an image where the LED corners are set and an image where the center LEDs are set. Figure 15.7 shows a frame with all 0s.

```
uint8_t frame[8][12] = {
  { 0, 0, 0, 0, 0, 0, 0, 0, 0, 0, 0, 0 },
  { 0, 0, 0, 0, 0, 0, 0, 0, 0, 0, 0, 0 },
  { 0, 0, 0, 0, 0, 0, 0, 0, 0, 0, 0, 0 },
  { 0, 0, 0, 0, 0, 0, 0, 0, 0, 0, 0, 0 },
  { 0, 0, 0, 0, 0, 0, 0, 0, 0, 0, 0, 0 },
  { 0, 0, 0, 0, 0, 0, 0, 0, 0, 0, 0, 0 },
  { 0, 0, 0, 0, 0, 0, 0, 0, 0, 0, 0, 0 },
  { 0, 0, 0, 0, 0, 0, 0, 0, 0, 0, 0, 0 }
};
```

Figure 15.7: Frame with all 0s.

Notice that you could have created a frame with all 0s with the following statement, but it may be useful to visualize the frame as an 8×12 matrix:

```
uint8_t frame[8][12] = {0};
```

The images **corners** and **center** are set in two functions as follows:

```
void corners()
{
        frame[0][0] = 1;
        frame[0][11] = 1;
        frame[7][0] = 1;
        frame[7][11] = 1;
}
```

and

```
void center()
{
        frame[3][5] = 1;
        frame[3][6] = 1;
        frame[4][5] = 1;
        frame[4][6] = 1;
}
```

The program calls the functions and then renders the images. Figure 15.8 shows the program listing (Program: **SimpleImages**).

```
//----------------------------------------------------------------------------
//                  CREATE SIMLE IMAGES
//                  ===================
//
// This program displays two simple images
//
// Author: Dogan Ibrahim
// File  : SimpleImages
// Date  : July, 2023
```

```
//-------------------------------------------------------------------------------
#include "Arduino_LED_Matrix.h"
ArduinoLEDMatrix matrix;

void setup()
{
  matrix.begin();
}

  //
  // Empty frame array
  //
  uint8_t frame[8][12] =
  {
  { 0, 0, 0, 0, 0, 0, 0, 0, 0, 0, 0, 0 },
  { 0, 0, 0, 0, 0, 0, 0, 0, 0, 0, 0, 0 },
  { 0, 0, 0, 0, 0, 0, 0, 0, 0, 0, 0, 0 },
  { 0, 0, 0, 0, 0, 0, 0, 0, 0, 0, 0, 0 },
  { 0, 0, 0, 0, 0, 0, 0, 0, 0, 0, 0, 0 },
  { 0, 0, 0, 0, 0, 0, 0, 0, 0, 0, 0, 0 },
  { 0, 0, 0, 0, 0, 0, 0, 0, 0, 0, 0, 0 },
  { 0, 0, 0, 0, 0, 0, 0, 0, 0, 0, 0, 0 }
  };

//
// Image corners
//
void corners()
{
       frame[0][0] = 1;
       frame[0][11] = 1;
       frame[7][0] = 1;
       frame[7][11] = 1;
}

//
// Image center
//
void center()
{
       frame[3][5] = 1;
       frame[3][6] = 1;
       frame[4][5] = 1;
       frame[4][6] = 1;
}
```

```
//
// Display the two images for 2 seconds each
//
void loop()
{
  corners();
  matrix.renderBitmap(frame, 8, 12);
  delay(2000);

  center();
  matrix.renderBitmap(frame, 8, 12);
  delay(2000);
}
```

Figure 15.8: Program: ***SimpleImages****.*

Figure 15.9 shows the images on the LED matrix.

Figure 15.9: Images.

15.2.3 Project 3: Using LED matrix 2 — creating a large + shape

Description: In this project, a simpler method is used to create an image on the LED matrix. Here again, a large **+** shape is created on the LED matrix.

Program listing: In this project, the Arduino **LED Matrix Tool** is used. This tool is available at the following link:

https://docs.arduino.cc/tutorials/uno-r4-wifi/led-matrix#animation-generation

Click on the given link to display the tool. This is shown in Figure 15.10. Use your mouse to create the required shape by clicking on the empty cells. Don't worry if you make a mistake, as you can erase the contents of a cell using the eraser tool. Once you have finished, click the button at the top right-hand corner of the screen (</>) to save the image. For example, give it the name **NewImage**. The file is given the extension **.h** and is saved in your Downloads folder with the contents as shown below:

```
const uint32_t NewImage[][4] = {
        {
                0x4004004,
                0xfff0400,
                0x40040040,
                66
        }
};
```

Since there is no animation in this project, the file has been modified as follows:

```
const uint32_t NewImage[] = {
                0x4004004,
                0xfff0400,
                0x40040040
};
```

Figure 15.10 Creating a shape using the LED Matrix Tool

Now you have to move this file to the folder where your sketch (program file) is and include it at the top of your program. You can display the image by calling the statement **matrix.loadFrame(NewImage)**. Figure 15.11 shows the program listing (Program: MatrixPlus2). The image is displayed in Figure 15.6.

```
//-----------------------------------------------------------------
//              CREATE a + IMAGE USING THE LED MATRIX TOOL
//              ==========================================
//
// This program displays the + image on the LED matrix
//
// Author: Dogan Ibrahim
// File  : MatrixPlus2
// Date  : July, 2023
//------------------------------------------------------------------
#include "Arduino_LED_Matrix.h"
#include "NewImage.h"
ArduinoLEDMatrix matrix;

void setup()
```

```
{
  matrix.begin();
}

//
// Display the image
//
void loop()
{
  matrix.loadFrame(NewImage);
  while(1);
}
```

Figure 15.11: Program: **MatrixPlus2**.

15.2.4 Project 4: Animation — displaying a word

Description: In this project, you create a simple animation. Here, the word **ELEKTOR** is displayed on the LED matrix.

Program listing: In this project, the **LED Matrix Tool** is used to create the letters to be animated. Figure 15.12 shows the completed images in the Tool. Notice that the numbers at the bottom of the images represent in milliseconds for how long the image will be displayed before moving to the next image. Here, each image is displayed for 1000 ms and the last image for 3000 ms. You can click the arrow icon at the top of the Tool to see the animation. Save the images under a name (e.g., MyElektor). The file **MyElektor.h** will be saved in your **Downloads** folder. Figure 15.13 shows the contents of this file.

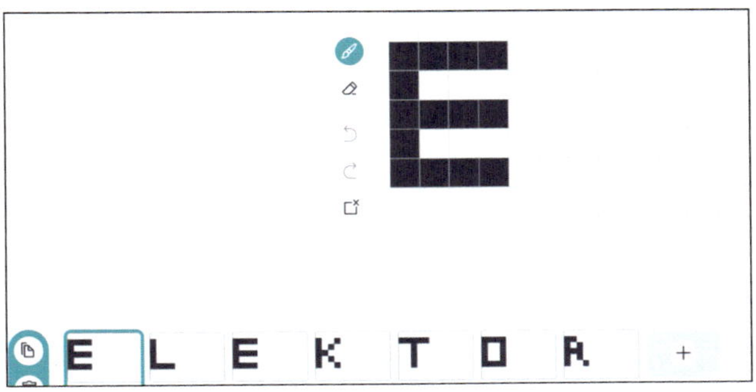

Figure 15.12: Completed images.

```
const uint32_t MyElektor[][4] = {
        {
                0xf0080,
                0xf00800f,
                0x0,
                1000
        },
```

```
        {
                0x80080,
                0x800800f,
                0x0,
                1000
        },
        {
                0xf0080,
                0xf00800f,
                0x0,
                1000
        },
        {
                0x900a0,
                0xc00a009,
                0x0,
                1000
        },
        {
                0xf8020,
                0x2002002,
                0x0,
                1000
        },
        {
                0xf0090,
                0x900900f,
                0x0,
                1000
        },
        {
                0xc00a0,
                0xe00a009,
                0x0,
                3000
        }
};
```

Figure 15.13: Contents of the animation file.

Figure 15.14 shows the program listing (Program: **Animate**). Function **Autoscroll()** has the frame scroll time in milliseconds so that there is no need to call the **next()** function. Function **play()** starts playing the loaded function.

```
//----------------------------------------------------------------------
//             ANIMATION - SCROLLING TEXT
//             ==========================
//
// This program scrolls the word ELEKTOR on the LED matrix
//
// Author: Dogan Ibrahim
// File  : Animate
// Date  : July, 2023
//-----------------------------------------------------------------------
#include "Arduino_LED_Matrix.h"
#include "MyElektor.h"
ArduinoLEDMatrix matrix;

void setup()
{
  matrix.begin();
  matrix.loadSequence(MyElektor);
  matrix.begin();
  matrix.play(true);
}

void loop()
{
}
```

Figure 15.14: Program: ***Animate****.*

15.3 Using the WiFi

Perhaps one of the key features of the Arduino Uno R4 WiFi board is its WiFi support. Before using the on-board WiFi, you have to specify the name (SSID) and password of your WiFi router. The header file **WiFiS3.h** must be included at the top of your program. In this section, a WiFi-based project is given to show how WiFi can be programmed.

Before going into the details of WiFi programming, it is worthwhile to review some of the WiFi programming terminology.

15.3.1 UDP and TCP

Communication over a WiFi link is in the form of a client and server, and sockets are used to send and receive data packets. The server side usually waits for a connection from the clients and once a connection is made two-way communication can start. Two protocols are mainly used for sending and receiving data packets over a WiFi link: UDP and TCP. TCP is a connection-based protocol that guarantees the delivery of packets. Packets are given sequence numbers and the receipt of all the packets is acknowledged to avoid them arriving in the wrong order. As a result of this confirmation, TCP is usually slow, but it is dependable as it guarantees the delivery of packets. UDP, on the other hand, is not connection-based.

Packets do not have sequence numbers and as a result of this, there is no guarantee that the packets will arrive at their destinations, or they may arrive in the wrong sequence. UDP has less overhead than TCP and as a result it is faster. Table 15.2 lists some of the differences between the TCP and UDP protocols.

TCP	UDP
Packets have sequence numbers and delivery of every packet is acknowledged	No delivery acknowledgement
Slow	Fast
No packet loss	Packets may be lost
Large overhead	Small overhead
Requires more resources	Requires less resources
Connection based	Not connection based
More difficult to program	Easier to program
Examples: HTTP, HTTPS, FTP	Examples: DNS, DHCP, Computer games

Table 15.2: TCP and UDP packet communications.

15.3.2 UDP communication

Figure 15.15 shows the UDP communication over a Wi-Fi link:

Server

1. Create a UDP socket.
2. Bind the socket to the server address.
3. Wait until the datagram packet arrives from the client.
4. Process the datagram packet.
5. Send a reply to the client or close the socket.
6. Go back to Step 3 (if not closed).

Client

1. Create a UDP socket (and optionally Bind).
2. Send a message to the server.
3. Wait until a response from the server is received.
4. Process reply.
5. Go back to step 2 or close the socket.

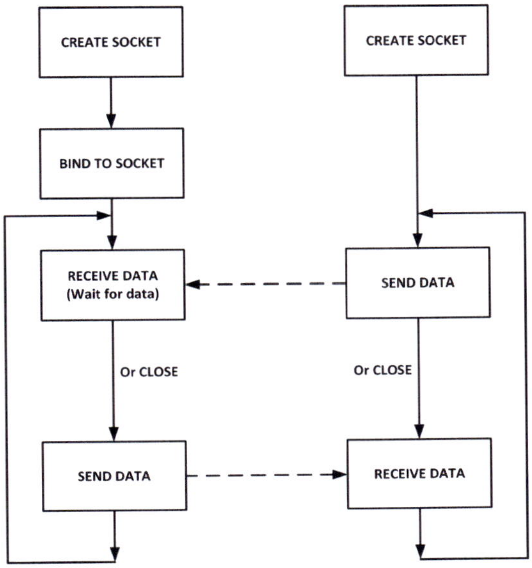

Figure 15.15: UDP communication.

15.3.3 TCP communication

Figure 15.16 shows the TCP communication over a Wi-Fi link:

Server

1. Create a UDP socket.
2. Bind the socket to the server address.
3. Listen for connections.
4. Accept connection.
5. Wait until the datagram packet arrives from the client.
6. Process the datagram packet.
7. Send a reply to the client or close the socket.
8. Go back to Step 3 (if not closed).

Client

1. Create a UDP socket.
2. Connect to the server.
3. Send a message to the server.
4. Wait until a response from the server is received.
5. Process reply.
6. Go back to step 2 or close the socket.

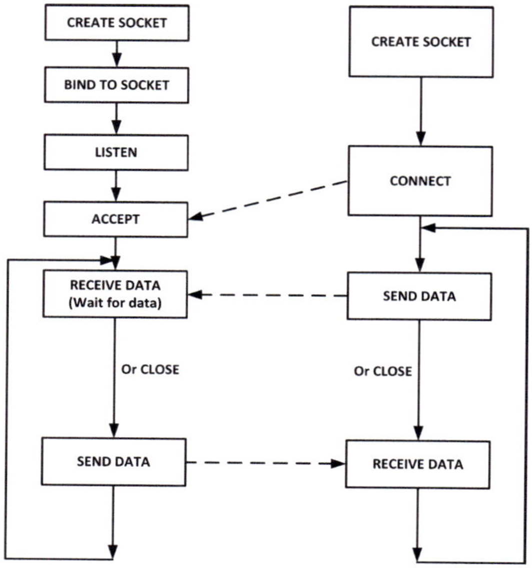

Figure 15.16: TCP communication.

15.3.4 Project 5: Controlling the Arduino Uno R4 WiFi on-board LED from a smartphone using UDP

Description: In this project, the on-board LED (at port 13) is turned ON and OFF by sending commands ON and OFF respectively from an Android smartphone. The aim of this project is to show how UDP can be programmed to send control messages over a network and control a device. In this project, Arduino is the server, and the smartphone is the client.

Block diagram: Figure 15.17 shows the block diagram of the project.

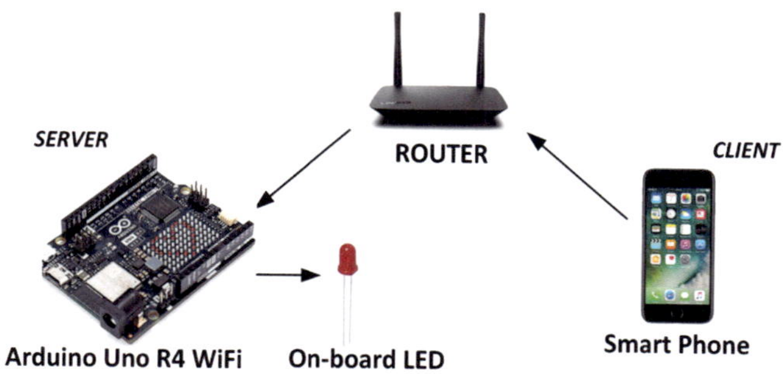

Figure 15.17: Block diagram of the project.

Program Listing: Figure 15.18 shows the program listing (program: **serverled**). A socket is created, and port 5000 is used. On-board LED is assigned to port 13, is configured as output, and is turned OFF at the beginning of the program. Arduino Uno R4 WiFi then connects to local WiFi and displays the SSID and IP address of the Arduino in the following format:

```
Attempting to connect to SSID: BTHomeSpot-XNH
SSID: BTHomeSpot-XNH
IP Address: 192.168.1.226
```

Where BTHomeSpot-XNH is the SSID name of the author's WiFi router and 192.168.1.226 is the IP address of the Arduino Uno R4 WiFi.

Inside the program loop, a packet is read over the WiFI and its contents are checked. If the packet contains the text **L=ON**, then the LED is turned ON and a message is displayed on the Serial Monitor. If, on the other hand, the text is **L=OFF** then the LED is turned OFF and a message is displayed on the Serial Monitor.

```
//----------------------------------------------------------------------
//             WiFi BASED REMOTE CONTROL
//             =========================
//
// In this program the on-board LED on the Arduino Uno R4 WiFi
// is controleld remotely from a smartphone over UDP link. The
// valid commands are: L=ON and L=OFF
//
// Author: Dogan Ibrahim
// File  : serverled
// Date  : July, 2023
//----------------------------------------------------------------------
#include "WiFiS3.h"
#include "WiFiUDP.h"

int LED = 13;                                  // On-board LED
int status = WL_IDLE_STATUS;

char ssid[] = "BTHomeSpot-XNH";                // Your WiFi SSID (name)
char pass[] = "49315vfg56b";                   // Your WiFi password
const int Port = 5000;                         // Port number used
char Packet[80];
WiFiUDP udp;

//
// Display the SSID and IP address of the WiFi connected to (OPTIONAL)
//
void printWifiStatus()
{
  Serial.print("SSID: ");
  Serial.println(WiFi.SSID());

  IPAddress ip = WiFi.localIP();
  Serial.print("IP Address: ");
```

```
    Serial.println(ip);
  }

  //
  // COnfigure the on-board LED, turn it OFF, configure WiFi
  //
  void setup()
  {
    pinMode(LED, OUTPUT);                              // LED is output
    digitalWrite(LED, LOW);                            // LED OFF at start
    Serial.begin(9600);
    delay(5000);

  //
  // Check for the WiFi module and stop if no WiFi module present
  //
    if (WiFi.status() == WL_NO_MODULE)
    {
      Serial.println("Communication with WiFi module failed!");
      while (true);
    }

  //
  // Attempt to connect to WiFi network
  //
    while (status != WL_CONNECTED)
    {
      Serial.print("Attempting to connect to SSID: ");
      Serial.println(ssid);
      status = WiFi.begin(ssid, pass);
      delay(10000);
    }
    printWifiStatus();                                 // Optional
    udp.begin(Port);                                   // listen to incoming packets
  }

  //
  // This is the main program loop. Inside the main program we read
  // UDP packets and then control the LED as requested. The format
  // of the control commands are:
  //
  // L=ON or L=OFF
  //
  // Any other commands are simply ignored by the program
  //
  void loop()
```

```
{
   int PacketSize = udp.parsePacket();
   if(PacketSize)
   {
     udp.read(Packet, PacketSize);

     if(Packet[1]=='=')
     {
        if(Packet[0]=='L')
        {
          if(Packet[2]=='O' && Packet[3]=='N')
          {
             digitalWrite(LED, HIGH);
             Serial.println("LED turned ON");
          }
          else if(Packet[2]=='O' && Packet[3]=='F' && Packet[4]=='F')
          {
             digitalWrite(LED, LOW);
             Serial.println("LED turned OFF");
          }
        }
     }
    }
}
```

Figure 15.18: Program: **serverled**.

The program can easily be tested using UDP apps on a smartphone. The author used the Android app called **UDP Sender *by hastarin*** (Figure 15.19). The server program is started, then the client is started. Figure 15.20 shows sending the **ON** command to turn ON the LED. Notice that the Arduino's IP address and the used port number are entered on the apps.

Figure 15.19: UDP Sender apps.

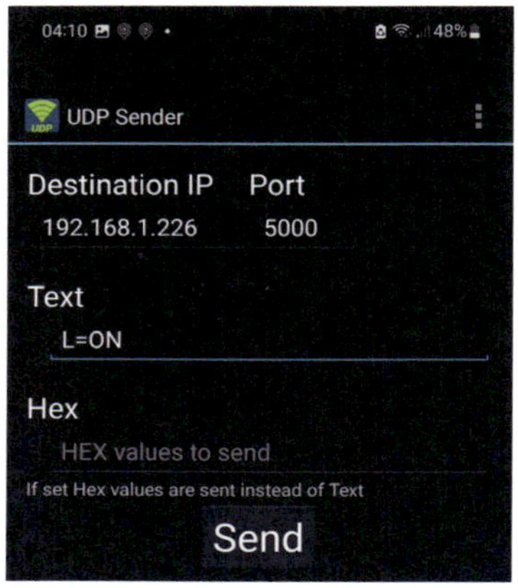

Figure 15.20: Command sent to turn ON the LED.

15.4 Bluetooth

The Arduino Uno R4 WiFi board has built-in Bluetooth 5 and Bluetooth LE hardware support, thus enabling communication with both old and new Bluetooth phones, tablets, and PCs.

Table 15.3 shows a comparison of classical Bluetooth and Bluetooth BLE. Classical Bluetooth has the following features:

- Multiple connections
- Up to 3 Mb/s data rate
- Up to 79 channels
- Continuous data streaming
- Device discovery
- Asynchronous data communication
- Master/slave Switch
- Adaptive frequency hopping
- Authentication and encryption
- Secure pairing
- Sniff mode
- Point-to-point network topology

Bluetooth BLE has the following features:

- Multiple connections
- Up to 2 Mb/s data rate
- Up to 40 channels
- Short burst data transmission
- Scanning
- Asynchronous data communication
- Data length extension
- Connection parameter update
- Point-to-point, broadcast, and mesh network topologies

	Bluetooth Low Energy (LE)	Bluetooth Basic Rate/ Enhanced Data Rate (BR/EDR)
Optimized For...	Short burst data transmission	Continuous data streaming
Frequency Band	2.4GHz ISM Band (2.402 – 2.480 GHz Utilized)	2.4GHz ISM Band (2.402 – 2.480 GHz Utilized)
Channels	40 channels with 2 MHz spacing (3 advertising channels/37 data channels)	79 channels with 1 MHz spacing
Channel Usage	Frequency-Hopping Spread Spectrum (FHSS)	Frequency-Hopping Spread Spectrum (FHSS)
Modulation	GFSK	GFSK, π/4 DQPSK, 8DPSK
Power Consumption	~0.01x to 0.5x of reference (depending on use case)	1 (reference value)
Data Rate	LE 2M PHY: 2 Mb/s LE 1M PHY: 1 Mb/s LE Coded PHY (S=2): 500 Kb/s LE Coded PHY (S=8): 125 Kb/s	EDR PHY (8DPSK): 3 Mb/s EDR PHY (π/4 DQPSK): 2 Mb/s BR PHY (GFSK): 1 Mb/s
Max Tx Power*	Class 1: 100 mW (+20 dBm) Class 1.5: 10 mW (+10 dbm) Class 2: 2.5 mW (+4 dBm) Class 3: 1 mW (0 dBm)	Class 1: 100 mW (+20 dBm) Class 2: 2.5 mW (+4 dBm) Class 3: 1 mW (0 dBm)
Network Topologies	Point-to-Point (including piconet) Broadcast Mesh	Point-to-Point (including piconet)

Table 15.3: Comparison of classical Bluetooth and Bluetooth BLE (Source: Bluetooth.com).

15.4.1 Bluetooth BLE

Bluetooth Low Energy (BLE) has been developed mainly for battery-operated short-distance communications and it consumes less power than classical Bluetooth. BLE is normally in sleep mode except when a connection is initiated and as a result of this feature, it is suitable for applications that require to exchange tiny amounts of data at regular intervals, such as in sending/receiving weather data, in healthcare, security, in home automation, in IoT applications, etc.

In Bluetooth BLE applications, you have a server and a client. In applications that you are interested in here, ESP32 acts as the server, and the PC or a mobile phone act as client, although this can be changed if required.

The server-client communication is normally done using a **point-to-point** protocol where the data exchange takes place as follows:

- The server makes itself known to nearby Bluetooth BLE client devices.
- The client devices scan their surroundings and when they find the server that they are looking for they establish a connection to the server.
- The client device listens for incoming data from the server.

Other modes of Bluetooth BLE communication are **Broadcast** and **Mesh**. In Broadcast mode, the server broadcasts data to a number of connected client devices. In Mesh mode, all the BLE devices are connected to each other, and they can exchange data. In this chapter, you will be developing projects using the point-to-point protocol.

15.4.2 Bluetooth BLE Software Model

Bluetooth BLE devices use the Generic Attributes Profile (GATT) which defines the way that two BLE devices can exchange data. GATT is a hierarchical data structure. As shown in Figure 15.21 GATT hierarchy consists of the Profile, Service, and Characteristics. At the top level, you have the Profile which consists of one or more Services. A Service consists of one or more Characteristics, and they can make references to other services. A Characteristic contains the actual data and consists of Properties, Values, and Descriptors.

Every Service, Characteristic, and Descriptor in a Profile has unique 16-byte (128-bits) Universally Unique identifiers (UUID) which identify a particular service provided by a BLE device. The **Bluetooth Special Interest Group** at:

https://www.bluetooth.com/specifications/gatt/services

gives lists of shortened UUIDs. For example, the **Battery Service** (used in portable battery-operated BLE devices to indicate the current battery level) has the Uniform Type Identifier: **org.bluetooth.service.battery_service** and its assigned number is **0x180F**. Using unique identifiers, any BLE device can find out the battery level, regardless of the manufacturer. Looking at the **Battery Service** you can see that **Battery Level** is a characteristic of this service and its UUID is 0x2A19 (see web site: **org.bluetooth.characteristic.battery_level**). The **Characteristic Descriptor** and **Characteristic Presentation Format** are the descriptors of this service and their UUIDs are 0x2902 and 0x2904, respectively. As an example, Figure 15.22 shows the GATT data structure for the **Battery Service**. If your application requires its own UUID, you can generate it from the following website:

https://www.uuidgenerator.net/

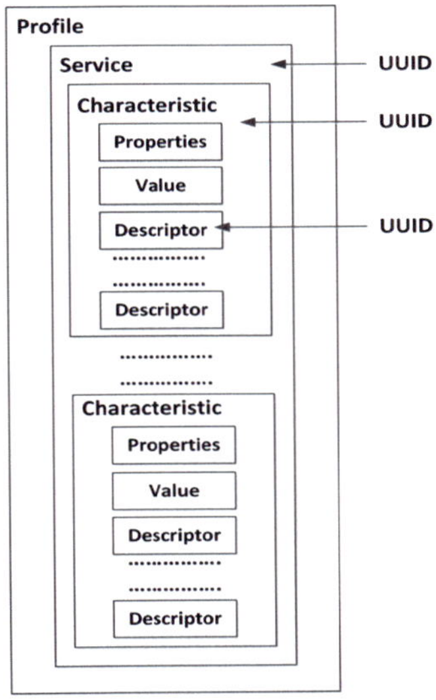

Figure 15.21: The GATT hierarchy.

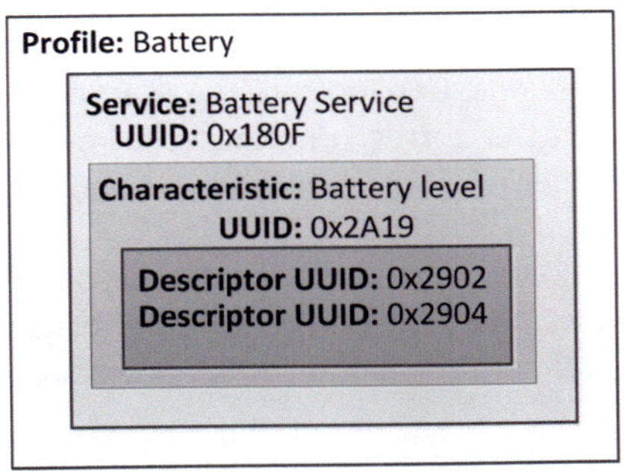

Figure 15.22: GATT data structure for the Battery Service.

Unfortunately, at the time of drafting this book the Radio Module (e.g., Bluetooth hardware) on the Arduino Uno R4 WiFi development board needed a firmware update to version 0.2.0 before Bluetooth could be used. Also, the Arduino BLE library was in Beta version, i.e., still under development. Interested readers can load the new firmware and the Arduino BLE library from the following link:

https://forum.arduino.cc/t/radio-module-firmware-version-0-2-0-is-now-available/1147361

The author has upgraded the firmware on his development board and also installed the Arduino BLE library successfully by following the instructions given in the above link. He has also tested some Bluetooth BLE-based programs successfully after the firmware upgrade and library installation.

Chapter 16 • Serial Communications

16.1 Overview

Serial communication is used in many applications and can be implemented in software or hardware. Hardware-based serial communication is usually done using a UART and this form of serial communication is not just more reliable than software-based, but also faster. Serial communication is normally done using 3 wires: Transmit (TX), Receive (RX), and Ground. In a typical application, the TX and RX pins are crossed over. i.e., the TX and RX pins of serial device 1 are connected to RX and TX pins of serial device 2.

One of the most important parameters of serial transmission is the baud rate. This is basically the number of bits transferred in one second. The baud rate can take values of 9600, 19200, 38400, etc. It is important that both devices have the same baud rate, and otherwise, they cannot communicate serially. Data is transferred in a frame. The most commonly used frame is: 1 start bit, 8 data bits, no parity bit, 1 stop bit, making a total of 10 bits. Therefore, for example, at a baud rate of 9600, 960 characters (or bytes) are transferred every second. The (optional) parity bit is used for error checking and can be odd or even. If used, the parity bit is added to the end of the data before the stop bit. With odd-parity communication, the parity bit is set or reset so that the number of 1s in the data is odd. Similarly, with even-parity communication, the number of 1s in the data is even. Normally, the line is at logic 1 and goes to logic 0 to indicate the start of the communication. The data bits are then sent serially with the correct timings, followed by the stop bit, which is the low-to-high transition of the line.

One advantage of using serial communication instead of parallel is that only 3 wires are used instead of 8 or more wires. Also, the communication distance is much longer with serial communication. Most serial communication wires nowadays carry +5 V or GND.

Traditionally, long-distance serial communication is implemented based on the RS232 protocol where the voltage levels are ±12 V, with −12 V representing Mark (or 1) and +12 V representing SPACE (or 0). Special ICs are used to translate the +5 V and GND voltages to ±12 V.

In this chapter, an example project is given to show how hardware-based serial communication can be used with the Arduino Uno R4 Minima.

The UNO R4 Minima board features two separate hardware serial ports.

- One port is exposed via USB-C
- One port is exposed via RX/TX pins.

This is one of the few things that are distinctly different from Uno R3 to Uno R4, as the former only features one hardware serial port that's connected to **both** the USB port and the RX/TX pins on the board.

The Arduino Uno supports many library functions for serial communication. Some of the commonly used ones are given below.

begin()	-	start serial communication
available()	-	check if serial data is available
read()	-	read a byte from the serial port
readBytesUntil()	-	read until a terminator character is detected
readString()	-	read characters into a string
readStringUntil()	-	read characters until a terminator character
print()	-	send data to the serial port
println()	-	send data terminated with a new line to the serial port
setTimeout()	-	specify the maximum time (in ms) to wait for serial data

See: https://www.arduino.cc/reference/en/language/functions/communication/serial/ for detailed information on all the supported functions

16.2 Project 1: Receiving ambient temperature from an Arduino Uno R3

Description: In this project, an Arduino Uno R3 reads the ambient temperature and sends it at a 1-second rate over a serial line to an Arduino Uno R4 Minima where it is displayed on the Serial Monitor. The aim of this project is to show how serial communication can be used.

Block diagram: Figure 16.1 shows the block diagram of the project. In this project, an LM35-type analog temperature sensor is used.

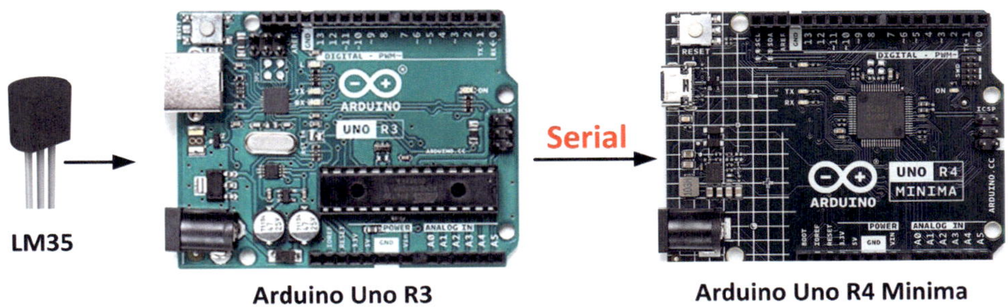

Figure 16.1: Block diagram of the project.

Circuit diagram: As shown in Figure 16.2, LM35 is connected to port pin A5 of the Arduino Uno R3. The serial TX pin (pin 1) of the Arduino Uno R3 is connected to the RX pin (pin 0) of the Arduino Uno R4 Minima. Additionally, the Ground pins are connected together.

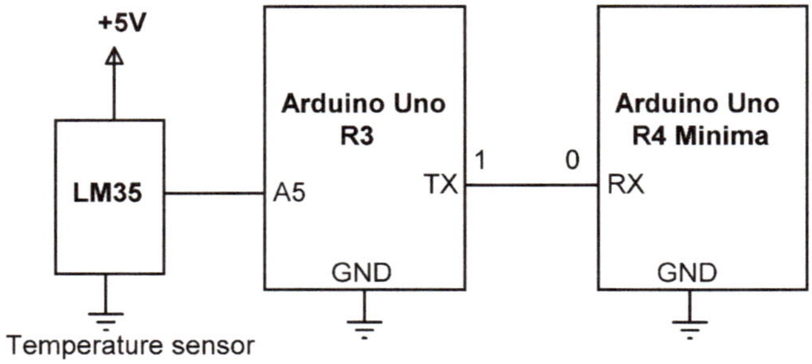

Figure 16.2: Circuit diagram of the project.

Arduino Uno R3 program listing: Figure 16.3 shows the Arduino Uno R3 program listing (Program: **SendTemp**). Software serial library is used for the Arduino Uno R3. Pins 2 and 3 are configured as serial RX and TX respectively. The baud rate is set to 9600. Inside the main program loop, the temperature is read from analog port A5 and is converted to degrees C. It is then converted into a string in variable **Temp** and terminated with a "#" character. The resulting data is sent over the serial line every second.

```
/*****************************************************
 *  SEND TEMPERATURE READINGS OVER SERIAL LINE
 *  ==========================================
 * This program sends he LM35 temperatur ereadings
 * to serial line every second
 *
 * Author : Dogan Ibrahim
 * Program: SendTemp
 * Date   : July, 2023
 ****************************************************/
#include <SoftwareSerial.h>

#define rxPin 2
#define txPin 3
SoftwareSerial MySerial(rxPin, txPin);

int LM35 = A5;
int TempValue = 0;
float mV;
int T;

void setup()
{
  MySerial.begin(9600);
}

//
```

```
// Read the temperature and send over serial line
//
void loop()
{
  TempValue = analogRead(LM35);          // Read temp
  mV = 5000.0 * TempValue / 1024.0;      // in mV
  T = int(mV / 10);                      // In degrees C
  String Temp = String(T) + "#";         // Terminate
  MySerial.print(Temp);                  // Send
  delay(1000);
}
```

Figure 16.3: Program: **SensTemp**.

Arduino Uno R4 Minima program listing: Figure 16.4 shows the Arduino Uno R4 Minima program listing (Program: **RcvTemp**). Hardware UART is used in this program to receive the temperature data. Function **readStringUntil()** is used to read the temperature until the terminator '#' key is detected. The temperature is displayed on the Serial Monitor as shown in Figure 16.5.

```
//----------------------------------------------------------------------
//             RECEIVE TEMPERATURE DATA OVER SERIAL LINE
//             =========================================
//
// In this program the temperature sent by the Arduino Uno R3 is read
// over the serial line (UART) and is displayed on the Serial Monitpr
//
// Author: Dogan Ibrahim
// File  : RcvTemp
// Date  : July, 2023
//-----------------------------------------------------------------------
const char Terminator = '#';
void setup()
{
  Serial1.begin(9600);                   // UART line
  Serial.begin(9600);                    // Serial Monitor
}


void loop()
{
 if(Serial1.available())
 {
   String T = Serial1.readStringUntil(Terminator);
   Serial.print("Temperature = ");
```

```
    Serial.println(T);
  }
}
```

Figure 16.4: Program: **RcvTemp**.

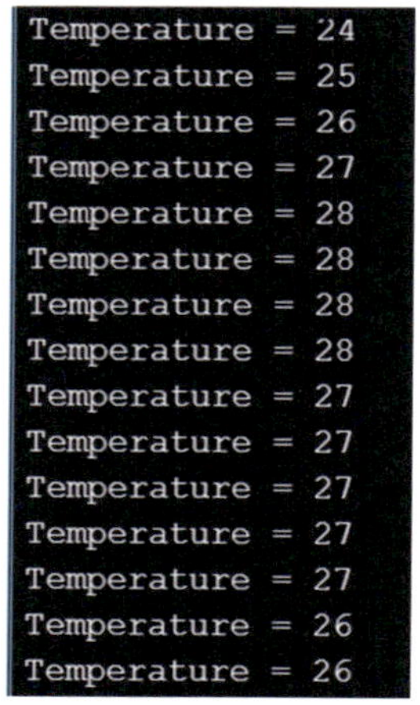

Figure 16.5: Temperature displayed on Serial Monitor.

Chapter 17 • Using an Arduino Uno Simulator

17.1 Why simulation?

Among the most salient features of engineering courses is its strong dependency on laboratory work. Students learn complex engineering theories in classes and then apply these theories to practice by carrying out experiments in laboratories. This is true for all engineering courses, whether chemical engineering, mechanical engineering, electronic engineering, etc. For example, electronic engineering students learn the architecture and programming of microcontrollers in class. They then carry out experiments in laboratories using real microcontroller hardware and programming tools such as assemblers, compilers, debuggers, and so on.

Although real laboratory experiments are very useful, they have some problems associated with them:

- Investment in real electronic components and instruments can be costly. A large number of identical instruments is usually required for a class of students and this raises the cost.
- The characteristics of electronic components and electronic equipment can change with ageing, wear and tear, and poor storage conditions.
- It is not always easy to find the required components and students may have to wait long before they can develop and test their projects.
- Real components can easily be damaged by improper use, for example by applying large voltages, or by passing large currents, or short-circuiting.
- Students can get electric shocks in laboratories by not following the safety regulations. Thus, an instructor must always be present in a laboratory to make sure that students connect the components correctly and follow the safety rules.
- Laboratory instruments usually need calibration from time to time and this can be costly as well as inconvenient.
- Students enrolled in distance learning courses may not be able to attend laboratories. This was also the case during the recent Covid-19 pandemic.

Computer simulation is an alternative to carrying out experiments using real hardware. A simulator is basically a computer program used to predict the behavior of a real circuit. Software models of real components like the Arduino Uno board, resistors, LEDs, buzzers, etc., are used in a simulator program. Typically, in a simulation, students run the simulator program and pick the required components and virtual instruments from a software library. For example, the value of a resistor can be changed with the click of a button. Then, they connect the required sensors and components to the target processor (e.g. an Arduino Uno) using the provided software tools. The simulation process is then activated and the response of the circuit is observed, printed, or plotted.

Although simulation is an alternative and an invaluable tool in designing and developing electronic projects, it has the following advantages and disadvantages:

+ Any component with whatever cost can be modeled and simulated using a simulator. Virtual instruments are computer programs, hence there are no cost issues.

– The simulation does not usually take into consideration any component tolerances, aging, or temperature effects. Users may come to think that all components are ideal at all times.

+ Virtual instruments and components used in a simulation can not be damaged by incorect connections or by applying excessive voltages or currents.

– Simulation results may not be accurate at very high frequencies. For example, concepts like the skin effect in inductors is not normally considered in a high-frequency simulation.

+ There are no calibration processes associated with virtual instruments. They are available at all times and always operate with the same specifications.

– Simulation allows measurements of internal currents and voltages that in many cases can be virtually impossible to access while using real components.

+ Simulation is easily embedded into distance education courses. Students can be supplied with copies of the simulator program, or they can be given access codes to use a simulator over the web (e.g. TINACloud). Experiments can then be carried out at the users' places of studying at times convenient to them.

– Simulation programs constantly evolve as new components are introduced by manufacturers. This requires updating the software from time to time.

In a typical Arduino Uno simulation exercise, students can graphically connect LEDs to the Arduino Uno, and then write a program to turn these LEDs ON and OFF. The behavior of the project can then be observed graphically.

There are several freely available Arduino Uno simulator software programs available on the Internet (e.g. Tinkercad, Wokwi, PICSimLab, Tina, SimulIDE, UnoArduSim, Virtronics, Proteus VSM, Virtual Breadboard, etc). In this chapter, you will be learning to use a simulator software called the **Wokwi**. This software can be used to simulate the Arduino Uno, ESP32, and STM32 processors.

17.2 The Wokwi simulator

This simulator is available free of charge at can be used from a web browser at the following link:

https://wokwi.com/

Perhaps the easiest way to learn to use this simulator is to carry out an example.

17.2.1 Project 1: A simple project simulation — flashing LED

In this project, you connect an LED through a 1-kΩ current-limiting resistor to port 7 of the Arduino Uno. A program will then be written to flash the LED every second.

The steps to simulate this circuit is as follows:

- Start the Wokwi simulator from the link given above.
- Scroll down and select **Arduino Uno** under the heading **Start from Scratch**.
- On the left-hand side, you will be presented with an Arduino IDE template, and on the right-hand side, you will see an Arduino board (Figure 17.1).

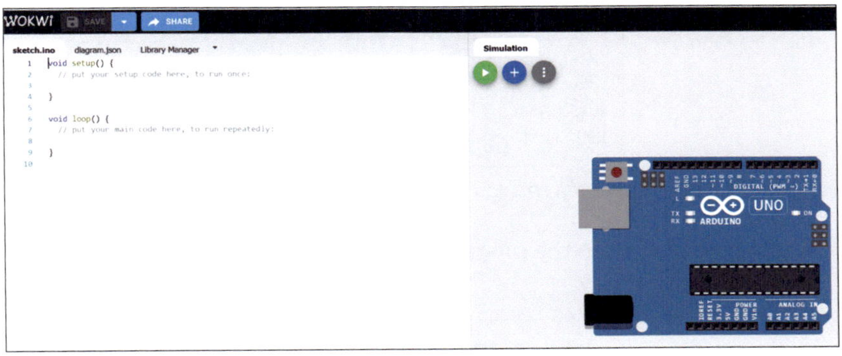

Figure 17.1: Arduino Uno simulator startup screen.

- You will now have to connect an LED and a resistor. Click on the Blue **+** sign and select **LED**. Also, click again and select **Resistor**. You should see an LED and a resistor at the top of the board (Figure 17.2).

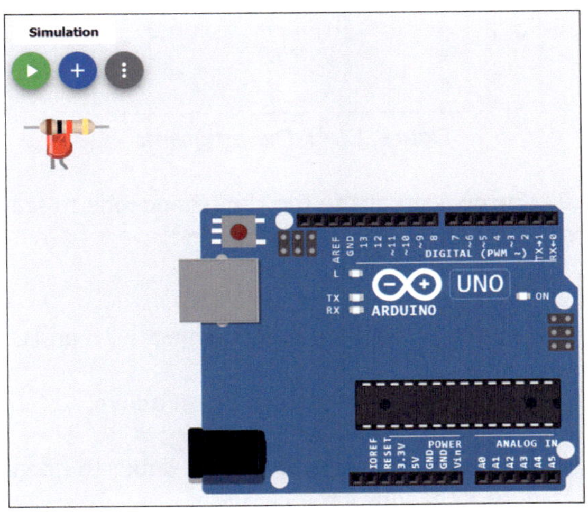

Figure 17.2: Select an LED and a resistor.

- Click on the LED and the resistor and move them. Connect the +ve pin of the LED (bent pin) to port pin 7 of the Uno, and the other pin to GND through the resistor as shown in Figure 17.3.

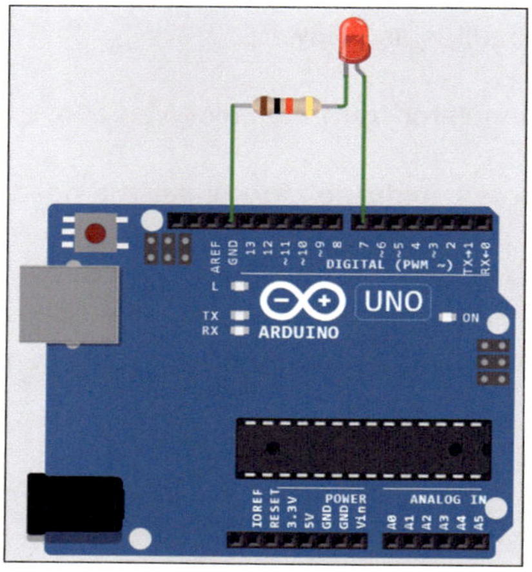

Figure 17.3: Connect the LED and the resistor to the Arduino Uno.

- You should now write the program. This is shown at the left-hand side of the screen in Figure 17.4.

sketch.ino ● diagram.json ● Libra

```
void setup() {
  pinMode(7, OUTPUT);
}

void loop() {
  digitalWrite(7, HIGH);
  delay(1000);
  digitalWrite(7, LOW);
  delay(1000);
}
```

Figure 17.4: The program.

- Click the Green arrow icon at the top right-hand side to start the simulation. You should see the LED flashing every second.

17.2.2 Project 2: Displaying text on LCD

In this project, you will select an already-designed project with an LCD. The steps are:

- Start the Wokwi simulator from the link given above.

- Scroll down and select **Arduino LCD 16x02** under the heading **Quick Start Templates.** Figure 17.5 shows the circuit.

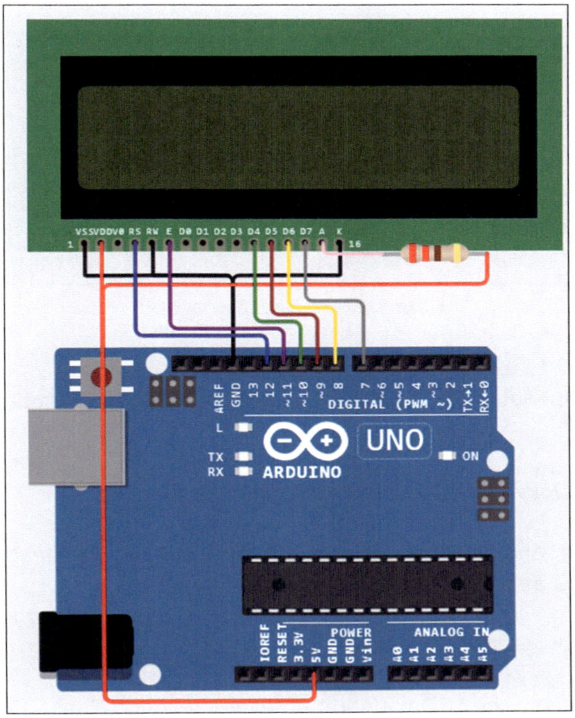

Figure 17.5: Arduino Uno with LCD.

- The program listing for this project is given on the left-hand side (see Figure 17.6).

lcd1602.ino   diagram.json   Library Manager

```
// LCD1602 to Arduino Uno connection example

#include <LiquidCrystal.h>

LiquidCrystal lcd(12, 11, 10, 9, 8, 7);

void setup() {
  lcd.begin(16, 2);
  // you can now interact with the LCD, e.g.:
  lcd.print("Hello World!");
}

void loop() {
  // ...
}
```

Figure 17.6: Program listing.

- Click the Green arrow icon to start the simulation, The text **Hello World** will be displayed on the LCD (Figure 17.7).

Figure 17.7: Displaying text.

17.2.3 Project 3: LCD seconds counter

In this project, the previous LCD hardware is used. Here, the LCD counts up every second. The steps are:

- Start the Wokwi simulator from the link given above.
- Scroll down and select **Arduino LCD 16x02** under the heading **Quick Start Templates** as in the previous project.
- Write the program as shown in Figure 17.8.
- You should see the LCD counting up every second (Figure 17.9).

lcd1602.ino ● diagram.json Library Manager

```
// LCD1602 Seconds counter example

#include <LiquidCrystal.h>
int count = 0;
char ary[10];

LiquidCrystal lcd(12, 11, 10, 9, 8, 7);

void setup() {
  lcd.begin(16, 2);
}

void loop() {
  lcd.clear();
  itoa(count, ary, 10);
  lcd.print(ary);
  count++;
  delay(1000);
}
```

Figure 17.8: Program listing.

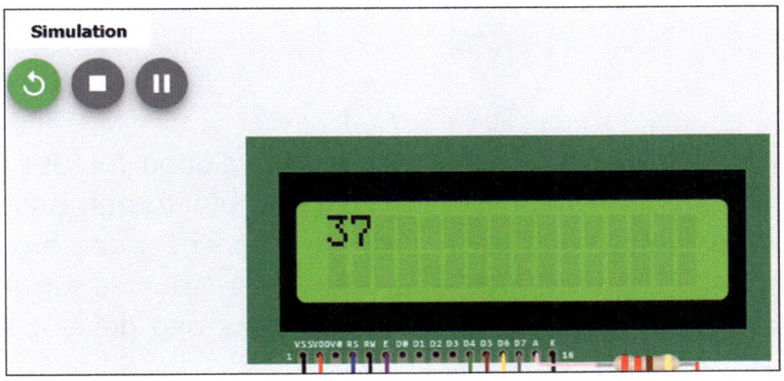

Figure 17.9: The LCD is counting up.

Suggestions: There are many built games and project templates available on the Wokwi website that you can try to make yourself familiar with simulating various sensors, displays, and actuators. Examples include "Simon" game, LED matrix tunnel, Arduino calculator, alarm clock, DHT22, TFT display, OLED display, keypad, servo, NTP clock, MQTT weather logger, joke machine, traffic controller, etc.

Chapter 18 • The CAN bus

18.1 Overview

The Controller Area Network (CAN) was originally developed for use in passenger cars. Today, CAN controllers are available from over 20 manufacturers, and CAN is finding applications in many other fields such as medical, aerospace, process control, automation, and so on. With the establishment of the CAN in Automation Association (CiA) in 1992, manufacturers and users have joined to exchange ideas and develop the CAN standards and specifications.

Fortunately, the new Arduino Uno R4 supports the CAN bus hardware and software. In this chapter, a brief introduction to the CAN bus will be given, followed by projects demoing the use of the CAN bus on the Arduino Uno R4. Detailed information on the CAN bus can be found on the Internet in many tutorials, datasheets, e-books, and application notes. You may find the book *The CAN Bus Companion* from Elektor very useful, especially as it comes with a CAN driver board.

18.2 The CAN bus

Figure 18.1 shows a CAN bus with three nodes. The bus is made up of a length of twisted-pair cable and is terminated at both ends with a resistor so that the bus characteristic resistance is 120 Ω. The two wires of the bus are termed CAN_H and CAN_L.

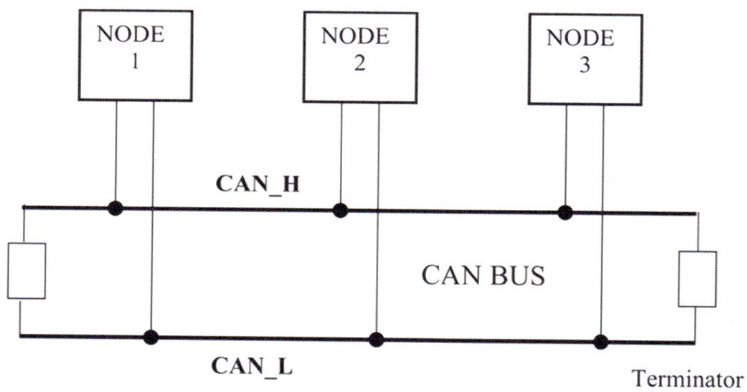

Figure 18.1: CAN bus with three nodes.

18.2.1 CAN bus termination

The bus is terminated to minimize signal reflections. Usually, a single 120-Ω resistor is connected at each end of the bus. Although the power rating of the chosen resistor is not particularly important, possible short circuits to power supplies on the bus should be considered when selecting a resistor power rating. A 0.25-watt, 5% tolerance resistor is generally used but it is recommended to use a higher power rating of 1 watt to avoid any damage to the bus due to conceivable transceiver short-circuits.

Although a single 120-Ω resistor is frequently used, in general, one of the following methods can be used to terminate the bus:

- Standard termination
- Split termination
- Biased split termination

The most commonly used termination method is the standard termination where a 120-Ω resistor is used at each end of the bus, shown in Figure 18.2.

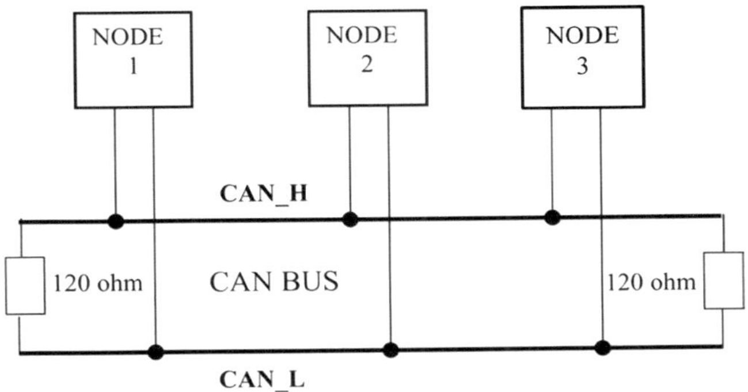

Figure 18.2: Standard bus termination.

Figure 18.3 shows split termination, which is gaining in popularity. In this method, two 60-Ω resistors and a capacitor are used at each end of the bus. The advantage of this method is that it eliminates high-frequency noise from the bus lines. Care must be taken to match the resistors so as not to reduce the effective noise immunity established on the bus. Typically, a 4.7-nF capacitor is chosen, which generates a response with a 3-dB roll off at approximately 1.1 Mbps.

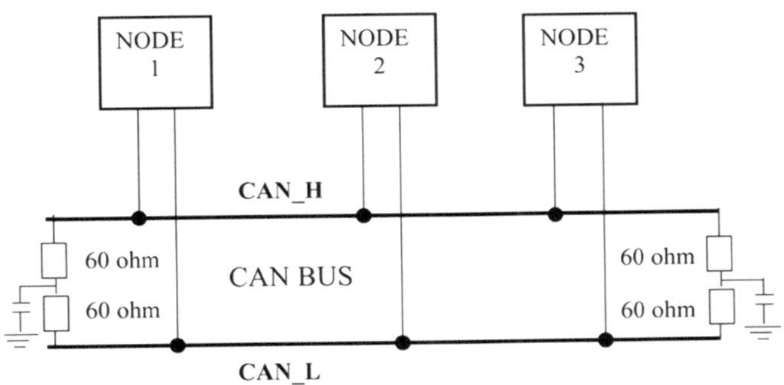

Figure 18.3: Split bus termination.

Figure 18.4 shows the biased split bus termination, where a voltage divider circuit and a capacitor are used at each end of the bus. As in split termination, the biased split termination increases the EMC performance of the bus.

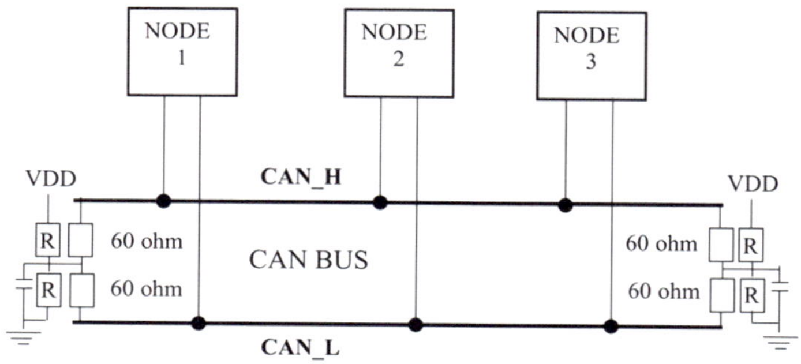

Figure 18.4: Biased split bus termination.

18.2.2 CAN bus data rate

ISO-11898 CAN specifies that a device on the bus must be able to drive a 40-m long cable at 1 Mb/s. In practice, a much higher bus length is achieved by lowering bus speed. Table 2.1 shows bus speed against bus length and nominal bit time. At 1000 kbps (1 µs nominal bit time) the maximum allowed bus length is 40 m, whereas at 10 kbps (100 µs nominal bit time) the maximum allowable bus length is increased to 6700 m.

Data rate (kbps)	Nominal bit time (µs)	Bus length (meters)
10	100	6700
20	50	3300
50	20	1300
125	8	530
250	4	270
500	2	130
1000	1	49

Table 18.1: Data rate against nominal bit time and maximum bus length.

A graph of the maximum data rate against the maximum allowed bus length is shown in Figure 18.5.

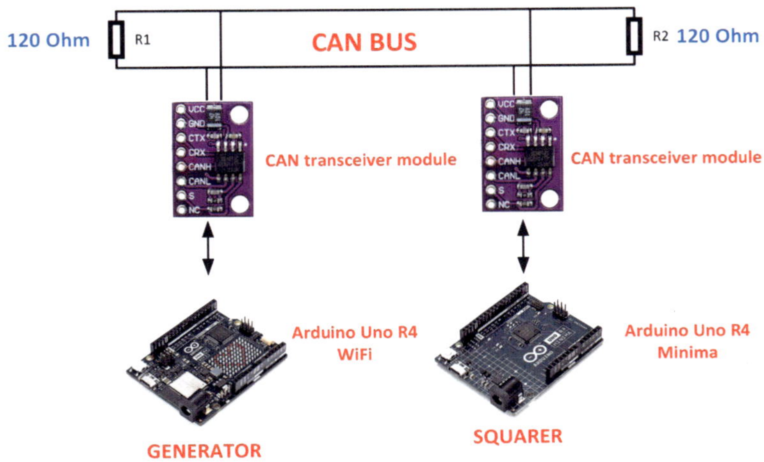

Figure 18.5: CAN bus data rate and maximum bus length.

In long cables, bus lines should be as close as possible to a straight line to keep possible reflections to a minimum. The cable length can be extended using a bridge device or repeater.

18.2.3 Cable stub length

In high data rate applications, the length of cable stubs and the distance between the nodes becomes an important factor. Since the stub lines are unterminated, signal reflections can develop on these lines, creating bus errors. At the maximum data rate of 1 Mbps, the length of the cable stubs (see Figure 18.6) should not be greater than 0.3 m, and the maximum node distance should be 40 m.

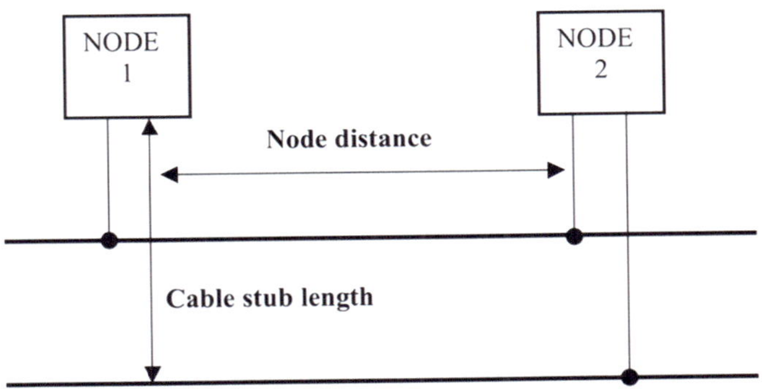

Figure 18.6: Cable stubs and node distances.

18.2.4 CAN Bus node

A CAN bus node consists of a CAN controller and a CAN transceiver (Figure 18.7). The CAN controller is connected to a microcontroller and the CAN transceiver is connected to the physical CAN bus. Some microcontrollers have built-in CAN controller circuitry (e.g. ARM) and thus a CAN node can be created using an external CAN transceiver chip. If the microcontroller does not have built-in CAN controller circuitry, then an external CAN controller and transceiver chips will be required to create a CAN node. CAN modules are available with built-in controller and transceiver chips. Such modules usually communicate with the microcontroller using the SPI bus.

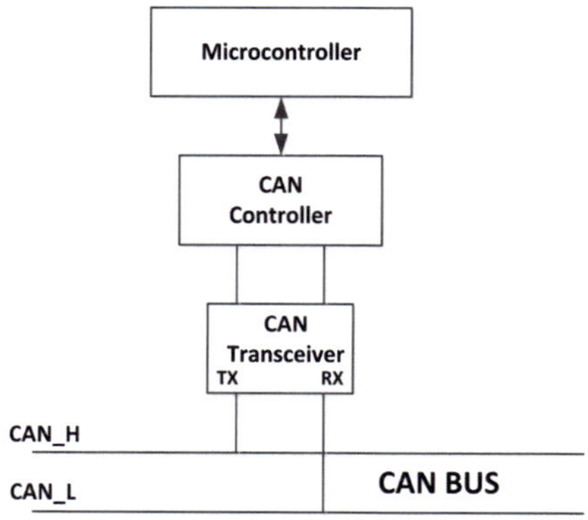

Figure 18.7: CAN bus node.

18.2.5 CAN bus signal levels

The data on the CAN bus is differential, and the bus specifies two logical states: dominant and recessive. Figure 18.8 shows the state of signals on the bus (see document *ISO-11898-4* for more details).

The recessive state is logic "1" and at this state, the differential voltage on the bus (i.e., V_{diff} = CAN_ H – CAN_L) is around 0 V (ideally CAN_H = CAN_L = 2.5 V). In practice, the recessive differential output voltage is less than 0.05 V at a bus transmitter output device.

The dominant state is logic "0" and at this state, the differential voltage on the bus (i.e. V_{diff} = CAN_ H – CAN_L) is around 2 V (ideally CAN_H = 3.5 V and CAN_L = 1.5 V). In practice, the dominant differential output voltage is between 1.5 V and 3.0 V.

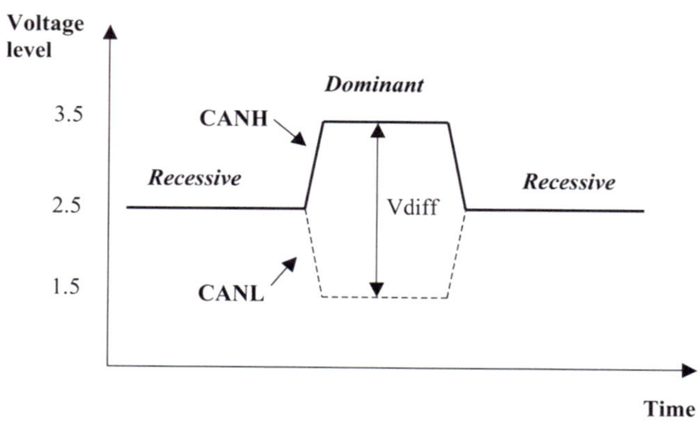

Figure 18.8: CAN bus signal levels.

18.2.6 CAN_H voltage

The CAN_H voltage should be half of the 5 V during times when no message is transmitted (or when the CAN bus is in a recessive state). This voltage can be measured using a digital voltmeter (DVM) if it is already known that the CAN controller is not sending any messages. In order to measure the CAN_H voltage, disconnect the TX pin from the transceiver and connect a +5 V supply to the transceiver TX input to force the transceiver to the recessive state (see Figure 18.9). Connect one lead of the meter to the CAN_H pin of the transceiver, and the other pin of the meter to the ground pin of the transceiver. The measured voltage should be around 2.5 V.

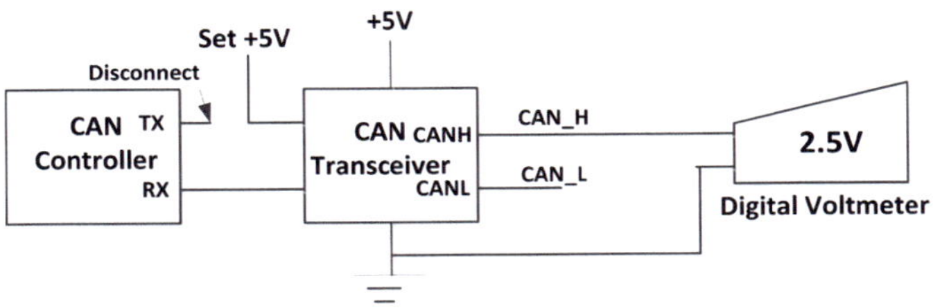

Figure 18.9: Measuring the CAN_H voltage.

18.2.7 The CAN_L voltage

The CAN_L voltage should be half of the available 5 V when no message is transmitted (or when the CAN bus is in a recessive state). As above, this voltage can be measured using a digital voltmeter. Use the same procedure as above but this time connect the voltmeter between CAN_L and the ground terminals (see Figure 18.10). The measured voltage should be around 2.5 V.

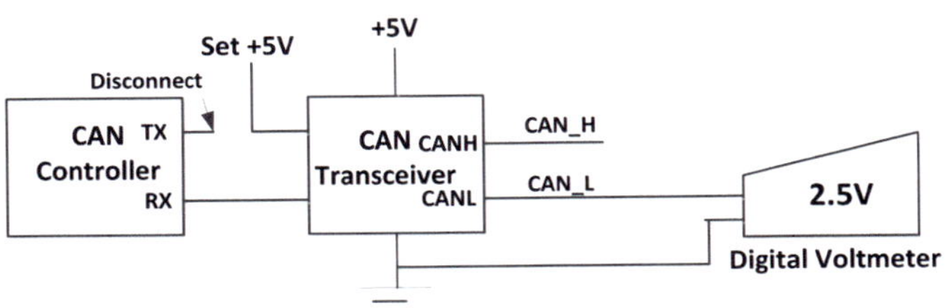

Figure 18.10: Measuring the CAN_L voltage.

18.2.8 Bus arbitration

When several bus nodes attempt to transmit at the same time, bus arbitration logic is applied to grant access to the bus. When there is arbitration on the bus, a dominant bit state always wins out over a recessive bit state.

18.2.9 Bus transceiver

The output of a CAN transceiver circuit is usually an "open-collector" (e.g. TTL logic) or "open-drain" (e.g. CMOS logic) configuration. When several such devices are connected to a bus, the net logic state of the bus is defined by the logical "AND" of the device outputs (also called the "Wired AND"). For example, if three devices are connected to the bus, the state of the bus will be logic "1" if (and only if) all the outputs of all the three devices are at logic "1", otherwise the bus will be at logic "0".

18.2.10 CAN connectors

Even though CAN is a two-wire network, in many cases a power signal and a Ground signal are added to the standard CAN connectors. As mentioned earlier, the actual wires used for the bus can either be unshielded twisted-pair (UTP) or shielded twisted-pair (STP). Shielded cables should be used in electrically noisy environments and when long bus cables are needed.

The standard CAN connector is a 9-pin D-type connector (DE-9) as shown in Figure 18.11. The pin configuration is shown in Table 2.2. Pin 2 and pin 7 are the CAN_L and CAN_H signals, respectively. Pin 3 and pin 9 are used as signal ground and signal power, respectively. Signal Ground and signal power pins can be useful when it is required to power remote devices. Care should be exercised to make sure that the current capacity of the cable used is not exceeded.

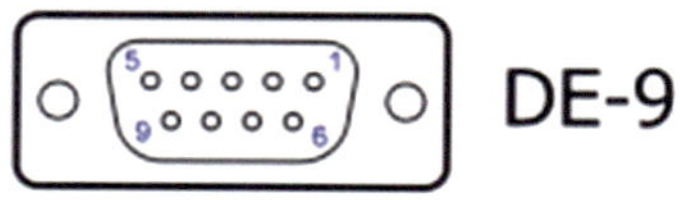

Figure 18.11: CAN D-connector.

9 Pin (male) D-Sub CAN Bus PinOut		
Pin #	**Signal Names**	**Signal Description**
1	Reserved	Upgrade Path
2	CAN_L	Dominant Low
3	CAN_GND	Ground
4	Reserved	Upgrade Path
5	CAN_SHLD	Shield, Optional
6	GND	Ground, Optional
7	CAN_H	Dominant High
8	Reserved	Upgrade Path
9	CAN_V+	Power, Optional

Table 18.2: CAN-D connector pin configuration.

Figure 18.12 shows a CAN-D type connector with a short CAN cable and terminating resistors.

Figure 18.12: CAN-D type connector with CAN bus cable.

Some companies use a 10-pin header to make connections to the bus. Table 2.3 gives the standard pin configuration for this type of connector.

10-Pin Header CAN Bus PinOut		
Pin #	**Signal Names**	**Signal Description**
1	Reserved	Upgrade Path
2	GND	Ground, Optional
3	CAN_L	Dominant Low
4	CAN_H	Dominant High
5	CAN_GND	Ground
6	Reserved	Upgrade Path
7	Reserved	Upgrade Path
8	CAN_V+	Power, Optional
6	Reserved	Upgrade Path
7	Reserved	Upgrade Path

Table 18.3: CAN 10-pin header pin configuration.

It is also common to use either RJ10 or RJ45-type connectors in CAN bus applications as shown in Figure 18.13. The pin configuration of this type of connector is shown in Table 2.4.

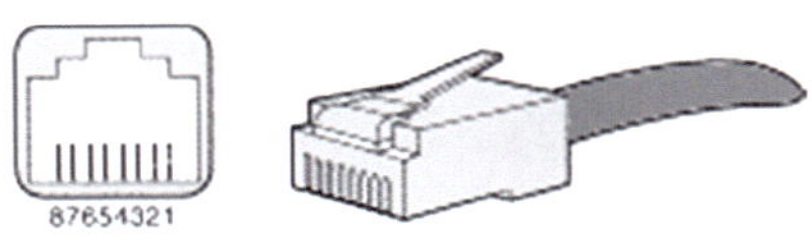

Figure 18.13: RJ10 or RJ45 type CAN connector.

RJ10, RJ45 CAN Bus PinOut			
RJ45 Pin #	**RJ10 Pin #**	**Signal Name**	**Signal Description**
1	2	CAN_H	Dominant High
2	3	CAN_L	Dominant Low
3	4	CAN_GND	Ground
4	-	Reserved	Upgrade Path
5	-	Reserved	Upgrade Path
6	-	CAN_SHLD	CAN Shield, Optional
7	-	CAN_GND	Ground
8	1	CAN_V+	Power, Optional

Table 18.4: RJ10 or RJ45 connector pin configuration.

18.3 Arduino Uno R4 CAN bus interface

The Arduino Uno R4 Minima has the following two pins that can be used for the CAN bus interface: D5 (RX) and D4 (TX). On the WiFi version, the CAN bus pins are: D13 (RX) and D10 (TX). Although there is CAN bus interface signals, it is required to connect CAN bus transceiver modules to these pins before connecting them to the physical CAN bus.

18.3.1 CAN bus transceivers

A CAN bus transceiver module is the physical interface between the Arduino Uno R4 CAN bus pins and the actual physical CAN bus cable. There are several transceiver modules available, and the one used in this chapter is the TJA1051. The TJA1051 (Figure 18.14) is a high-speed CAN transceiver module that provides an interface between a CAN protocol controller and the physical two-wire CAN bus.

The basic features of the TJA1051 are:

- Supply voltage 4.5–5.5 V
- ISO 11898-2:2016 and SAE J2284-1 to SAE J2284-5 compliant
- Suitable for 12 V and 24 V systems
- Low electromagnetic emission
- Supply current (1 mA in silent mode, 5 mA in bus recessive mode, 50 mA in bus dominant mode)

Figure 18.14: TJA1051 transceiver module.

The transceiver module has the following pins:

CANH	CAN_H bus interface
CANL	CAN_L bus interface
VCC	+5 V
GND	GND
CTX	Transmit data input
CRX	Receive data output (reads data from the bus)
S	Silent mode control (LOW in normal mode, HIGH in silent mode)

A CAN bus-based project is given in the following section to illustrate how the CAN bus feature of the Arduino Uno R4 can be used. The built-in Arduino_CAN library is used to communicate with other CAN devices.

18.4 Project 1: Arduino Uno R4 WiFi to Arduino Uno R4 Minima CAN bus communication

Description: In this project, you will have two Arduino Uno R4 boards. Arduino Uno R4 WiFi is named the **GENERATOR** and the Arduino Uno R4 Minima is named the **SQUARER** and both boards are connected on a CAN bus (i.e. there are two nodes called **GENERATOR** and **SQUARER**). The **GENERATOR** node will generate random integer numbers between 1 and 20 and send them to the **SQUARER** node, where the received numbers are squared and displayed on the Serial Monitor of the IDE. This process is repeated after a 3-second delay. The aim of this project is to show how two Arduino Uno R4s can be connected and communicate on a CAN bus.

Block diagram: Figure 18.15 shows the block diagram of the project. Here, the **GENERATOR** node is Arduino Uno R4 WiFi, and the **SQUARER** node is Arduino Uno R4 Minima. Two CAN bus transceiver modules, and the CAN bus wire are also shown.

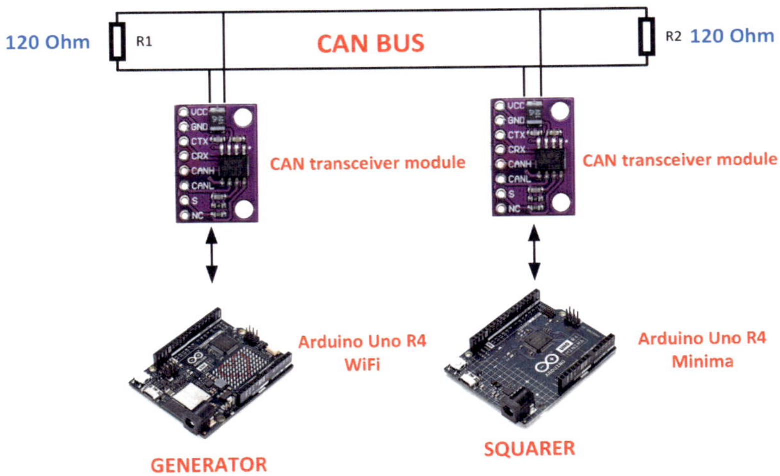

Figure 18.15: Block diagram of the project.

Note: The TJA1051 IC and the CAN bus transceiver modules and terminating resistors are not included in the kit. Also, only one (1) development board is included.

Circuit diagram: Figure 18.16 shows the circuit diagram of the project. Notice that the bus cable is terminated with two 120-Ω resistors. The CANH and CANL terminals of both transceivers are connected to the bus cable. The CTX and CRX pins of the transceiver module are connected to the Arduino Uno R4 WiFi D10 and D13 pins. Similarly, the CTX and CRX pins of the other transceiver are connected to the Arduino Uno R4 Minima D4 and D5 pins, respectively. In this project, about 1-meter length of twisted-pair cable was used as the CAN bus cable.

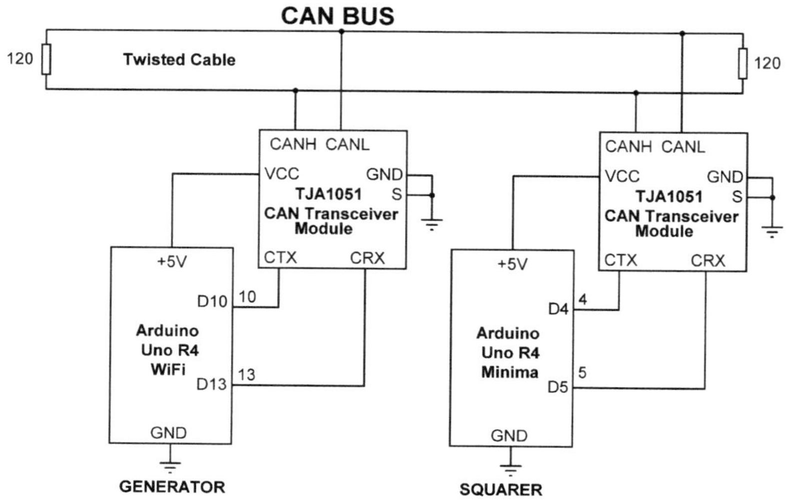

Figure 18.16: Circuit diagram of the project.

The **Arduino_CAN** library supports the following baud rates:

BR_125k, BR_250k, BR_500k, BR_1000k

The required baud rate must be specified during the initialization. For example, for 250 kbit/s baud rate, the required statement is: **CAN.begin(CanBitRate::BR_250k).**

Sending data
A **CanMsg** message object is created with the **CAN_ID**, size and data, and is sent over the bus using statement: **CAN.write()**

Receiving data
CAN.available() is used to check for data, and then if data is available it is read using the statement **CAN.read()**.

Node: GENERATOR program listing: Figure 18.17 shows the program listing (Program: **GENERATOR**). At the beginning of the program, the required header files are included at the beginning of the program. Then, array **msg_data** is initialized with 8 elements. This array will store the data to be sent over the CAN bus. **CAN_ID** is set to 0x20. Inside the **setup()** function, the CAN bus baud rate is set to 250 KB/s and CAN is started.

Inside the main program loop, a random number is generated between 1 and 20 and is sent over the CAN bus. Notice that only the first byte of the 8-byte data array **msg_data** is initialized. This process is repeated after 3 seconds of delay where a new random number is generated and sent to node **SQUARER**.

```
//------------------------------------------------------------------------
//              CAN BUS PROGRAM
//              ===============
//
// This program generates random numbers between 1 and 20 and sends them
// over the CAN bus to a node called SQUARER.The SQUARER node takes the
// square of these numbers and displays them on the Serial Monitor. CAN
// baud rate is set to 250 K.The process is repeated every 3 seconds
//
// Author: Dogan Ibrahim
// File  : GENERATOR
// Date  : July, 2023
//-------------------------------------------------------------------------
#include <Arduino_CAN.h>

static uint32_t const CAN_ID = 0x20;
uint8_t RandomNumber;
uint8_t msg_data[8];

void setup()
{
  Serial.begin(9600);
  delay(5000);

  if (!CAN.begin(CanBitRate::BR_250k))
  {
    Serial.println("CAN begin failed...");
    while(1);
  }
  else
    Serial.println("CAN begin success...");
}

void loop()
{
  RandomNumber = random(1, 21);                   // Random number
  Serial.print("Random number is: ");
  Serial.println(RandomNumber);
  msg_data[0] = RandomNumber;
  msg_data[1] = 0x00;
  msg_data[2] = 0x00;
  msg_data[3] = 0x00;
  msg_data[4] = 0x00;
  msg_data[5] = 0x00;
  msg_data[6] = 0x00;
  msg_data[7] = 0x00;
```

```
  CanMsg msg(CAN_ID, sizeof(msg_data), msg_data);

  int const rc = CAN.write(msg);
  if(rc < 0)
  {
    Serial.print("CAN write failed. Error code is: ");
    Serial.println(rc);
    while(1);
  }

  delay(3000);                              // Wait 3 secs and repeat
  }
```

Figure 18.17: Program: **GENERATOR***.*

Node: SQUARER program listing: Figure 18.18 shows the program listing (Program: **SQUARER**). The beginning and **setup()** function of this program is similar to Figure 18.17. Inside the main program loop, the program checks if there are any messages on the CAN bus and reads these messages. In this program, a message is an integer number between 1 and 20. The received number is stored in variable **Num** and its square is displayed on the Serial Monitor.

```
//----------------------------------------------------------------------
//              CAN BUS PROGRAM
//              ===============
//
// This program receives random numbers between 1 and 20 over the CAN bus
// and takes teh square of these numbers and then dislays the numbers on
// the Serial monitor
//
// Author: Dogan Ibrahim
// File  : SQUARER
// Date  : July, 2023
//----------------------------------------------------------------------
#include <Arduino_CAN.h>

static uint32_t const CAN_ID = 0x20;

void setup()
{
  Serial.begin(9600);
  delay(5000);

  if (!CAN.begin(CanBitRate::BR_250k))
  {
```

```
    Serial.println("CAN begin failed...");
    while(1);
  }
  else
    Serial.println("CAN begin success...");
}


void loop()
{
  if(CAN.available())
  {Serial.println("available");
    CanMsg const msg = CAN.read();
    int Num = msg.data[0];                      // Get the number
    Serial.print("Received number is: ");
    Serial.println(Num);                        // Received number
    Serial.print("Square is: ");
    Serial.println(Num * Num);                  // Square of number
    Serial.println("");
  }
}
```

Figure 18.18: Program: **SQUARER***.*

Testing the project: Construct the project as shown in the circuit diagram and power up both Arduino Unos. Start the Serial Monitor on the **SQUARER** node and you should see the received numbers and their squares displayed every 3 seconds. The sample output is shown in Figure 18.19.

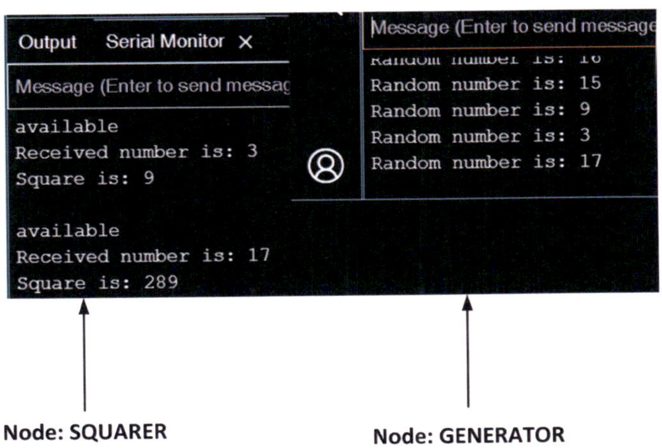

Figure 18.19: Sample output on Serial Monitor.

18.5 Project 2: Sending the temperature readings over the CAN bus

Description: As in the previous project, two Arduino Uno R4s are used. Node **GETTEMP** is an Uno R4 WiFi and it reads the ambient temperature. Node **DISPTEMP** is an Uno R4 Minima, and it displays the temperature on the Serial Monitor. The temperature data is sent every 3 seconds to node DISPTEMP. An analog temperature sensor (LM35) is used in this project.

Block diagram: Figure 18.20 shows the block diagram.

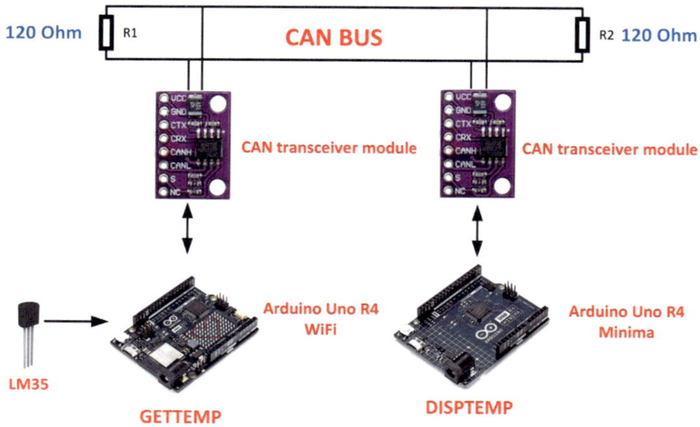

Figure 18.20: Block diagram of the project.

Circuit diagram: The circuit diagram is similar to Figure 18.16 and is shown in Figure 18.21. Analog temperature sensor LM35 is connected to analog input A5 of node **GETTEMP**.

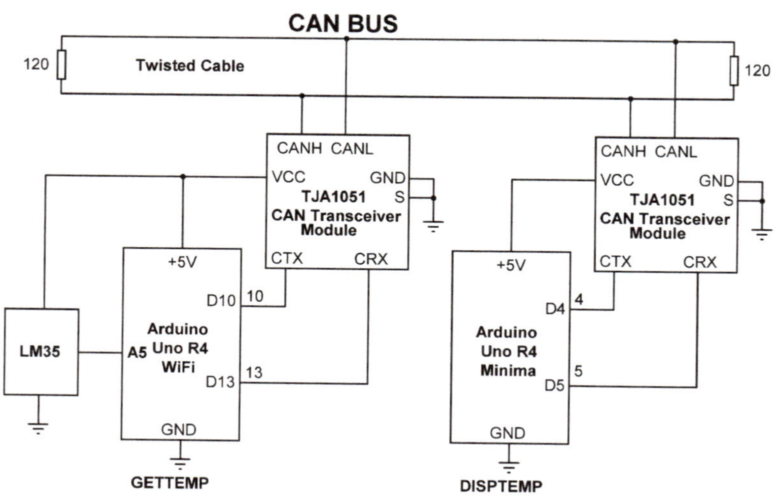

Figure 18.21: Circuit diagram of the project.

Node GETTEMP program listing: Figure 18.22 shows the program listing (Program: **GETTEMP**). The program is similar to Figure 18.17. Here, the temperature is read from analog port A5 and is sent over the CAN bus every 3 seconds. In this program, the default ADC resolution of 10 bits is used.

```
//------------------------------------------------------------------------
//             GET TEMPERATURE AND SEND OVER CAN BUS
//             =====================================
//
// This program reads the temperature from analog port A5 and sends over
// the CAN bus every 3 seconds
//
// Author: Dogan Ibrahim
// File  : GETTEMP
// Date  : July, 2023
//-------------------------------------------------------------------------
#include <Arduino_CAN.h>
int LM35= A5;
int val = 0;
float mV;
float conv = 5000.0 / 1024;

static uint32_t const CAN_ID = 0x20;
uint8_t msg_data[8];

void setup()
{
  Serial.begin(9600);
  delay(5000);

  if (!CAN.begin(CanBitRate::BR_250k))
  {
    Serial.println("CAN begin failed...");
    while(1);
  }
  else
    Serial.println("CAN begin success...");
}

void loop()
{
  val = analogRead(LM35);                    // Read raw temp
  mV = val * conv;                           // in mV
  int T = int(mV / 10);                      // in degrees C
  Serial.print("Temp = ");
  Serial.println(T);
```

```
  msg_data[0] = T;
  msg_data[1] = 0x00;
  msg_data[2] = 0x00;
  msg_data[3] = 0x00;
  msg_data[4] = 0x00;
  msg_data[5] = 0x00;
  msg_data[6] = 0x00;
  msg_data[7] = 0x00;

  CanMsg msg(CAN_ID, sizeof(msg_data), msg_data);

  int const rc = CAN.write(msg);
  if(rc < 0)
  {
    Serial.print("CAN write failed. Error code is: ");
    Serial.println(rc);
    while(1);
  }

  delay(3000);                                    // Wait 3 s
  }
```

Figure 18.22: Program: ***GETTEMP****.*

Node DISPTEMP program listing: Figure 18.23 shows the program listing (Program: **DISPTEMP**). The program is similar to Figure 18.18. Here, the temperature is displayed on the Serial Monitor.

```
//----------------------------------------------------------------------
//            DISPLAY TEMPERATURE
//            ===================
//
// This program receives temperature over the CAN bus and displays it
//
// Author: Dogan Ibrahim
// File  : DISPTEMP
// Date  : July, 2023
//----------------------------------------------------------------------
#include <Arduino_CAN.h>

static uint32_t const CAN_ID = 0x20;

void setup()
{
  Serial.begin(9600);
```

```
  delay(5000);

  if (!CAN.begin(CanBitRate::BR_250k))
  {
    Serial.println("CAN begin failed...");
    while(1);
  }
  else
    Serial.println("CAN begin success...");
}


void loop()
{
  if(CAN.available())
  {
    CanMsg const msg = CAN.read();
    int T = msg.data[0];                          // Get the temp
    Serial.print("Received temperature is: ");
    Serial.println(T);                            // Received temp
  }
}
```

Figure 18.23: Program: ***DISPTEMP****.*

Figure 18.24 shows the temperature displayed on node **DISPTEMP**. The project was constructed on a breadboard as shown in Figure 18.25.

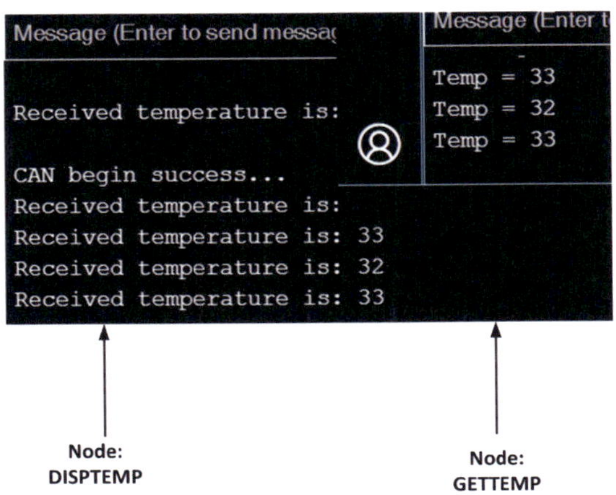

Figure 18.24: Example display on node ***DISPTEMP****.*

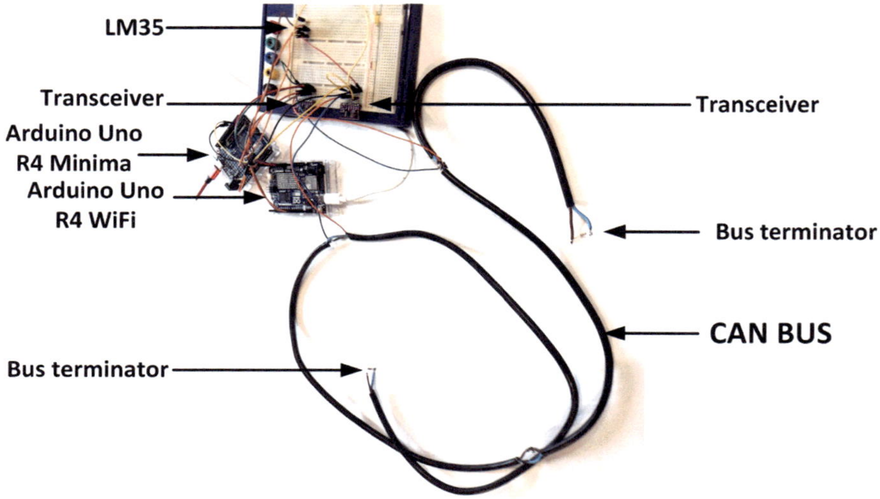

Figure 18.25: Project constructed on a breadboard.

Chapter 19 • Infrared Receiver and Remote Controller

19.1 Overview

Infrared (IR) communication is widely used in many applications such as TV/video remote control, motion sensors, infrared thermometers, etc. In this chapter, you will learn how to use the infrared receiver and transmitter modules supplied in the kit.

IR communication is widely used thanks to its ease of use and very low cost. When you press the button on your TV remote control, the IR transmitter LED turns ON and OFF and sends a modulated IR signal to the TV receiver. This signal is demodulated by the TV and the required command is executed. One of the common modulation techniques used for IR communication is called 38 kHz modulation. Using modulation prevents the signal conveyed from transmitter to receiver from being affected by ambient light. Many manufacturers use their own codes for their IR remote control systems, while others share their codes for compatibility.

19.2 The supplied infrared receiver

The infrared receiver supplied with the kit (Figure 19.1) is a photodiode and preamplifier that converts IR light into electrical signals.

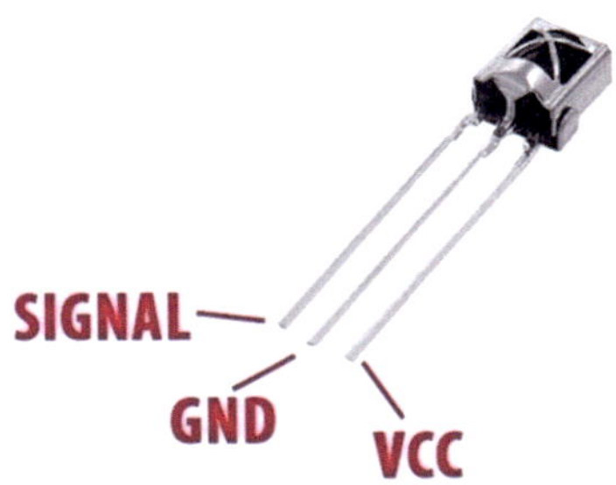

Figure 19.1: Infrared receiver supplied with the kit.

The IR receiver has 3 pins: GND (middle pin), VCC (right-hand pin facing the receiver glass), and the output pin (left-hand pin facing the receiver glass). This receiver operates at 38 kHz and works at 2.1 V to 6.5 V, with a supply current not exceeding 1.5 mA. The maximum reception distance is around 15 meters assuming a line of sight between the transmitter and the receiver.

19.3 The supplied infrared remote control transmitter unit

The supplied IR transmitter unit (Figure 19.2) is a keypad-based remote control unit operating with a CR2025-type coin battery.

Figure 19.2: Infrared remote control transmitter.

Before you can use the IR receiver and transmitter, you have to know the codes sent by the remote control unit when a button is pressed. Knowing this information, you can develop projects to control devices remotely from the remote control unit.

The popular Arduino Uno R3 **IRRemote** control library was not supported by the Arduino Uno R4 family at the time of drafting this book. Because of this, the author has developed a program from the first principles in order to do infrared-based remote control applications. It is important to understand how an infrared remote control system works before the operation of this program can be understood. This is explained in detail in the next section.

19.4 Operation of an infrared remote control system

There are many infrared-based remote control standards (NEC, Philips RC5, Philips RC6 SIRC, Sony, etc.) developed by various manufacturers. In this section, the popular NEC control protocol is explained since this is the protocol used by the author.

A typical infrared communication system requires an IR transmitter and an IR receiver. The transmitter looks like a standard LED, except it produces light in the IR spectrum instead of the visible spectrum. The IR receiver, on the other hand, is a photodiode and pre-amplifier that converts the IR light into an electrical signal that can be used by a digital processor. As mentioned earlier, IR light is modulated so that it is not affected by environmental light in the form of noise. In IR modulation, an encoder on the IR controller unit (remote control handheld unit) converts a binary signal into a modulated electrical signal, which is sent to the transmitting IR LED, thus emitting an infrared light signal. The receiver then demodulates this IR light signal and converts it back to binary so that it can be used by a processor. The carrier frequency used by most transmitters is 38 kHz, because it is rare in nature and thus can be distinguished from ambient noise. This way the IR receiver will know that the 38 kHz signal was sent from the transmitter and not picked up from the surrounding environment. Figure 19.3 shows as a block diagram how the infrared remote control system works.

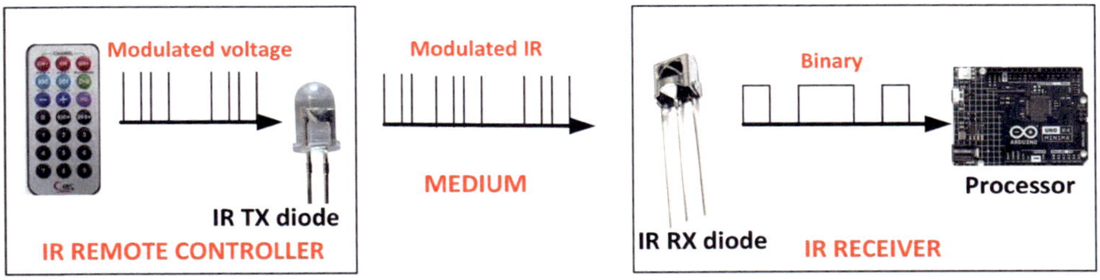

Figure 19.3: Infrared remote control system.

The NEC IR transmission protocol uses pulse encoding of the message bits. Each pulse burst is 562.5 µs in length, at a carrier frequency of 38 kHz (26.3 µs). Logic bits are transmitted as follows:

- Logic '0' — a 562.5-µs pulse burst followed by a 562.5-µs space, with a total transmit time of 1.125 ms.
- Logic '1' — a 562.5-µs pulse burst followed by a 1.6875-ms space, with a total transmit time of 2.25 ms.

Figure 19.4 shows the data transmission protocol. The image was taken from the site: https://techdocs.altium.com/display/FPGA/NEC+Infrared+Transmission+Protocol).

When a key is pressed on the remote controller, the message transmitted consists of the following, in the order given:

- a 9-ms leading pulse burst (16 times the pulse burst length used for a logical data bit)
- a 4.5-ms space
- the 8-bit address for the receiving device
- the 8-bit logical inverse of the address
- the 8-bit command
- the 8-bit logical inverse of the command
- a final 562.5-µs pulse burst to signify the end of message transmission.

Notice from image that it takes:

- 27 ms to transmit both the 16 bits for the address (address + inverse) and the 16 bits for the command (command + inverse). This comes from each of the 16-bit blocks ultimately containing eight '0's and eight '1's — giving (8 * 1.125 ms) + (8 * 2.25 ms).
- 67.5 ms to fully transmit the message frame (discounting the final 562.5 µs pulse burst that signifies the end of the message).

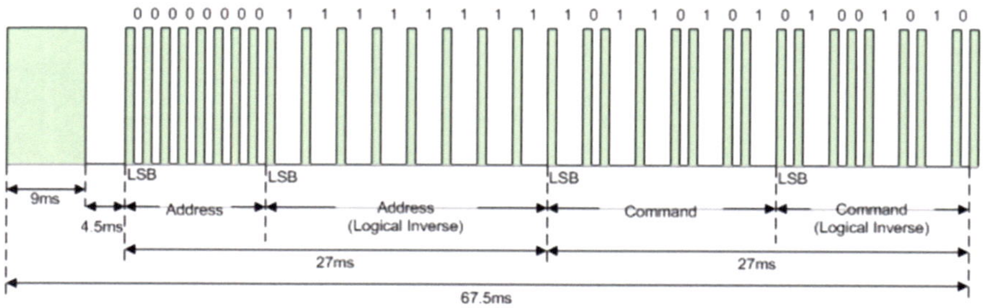

Fgure 19.4: NEC infrared data transmission.

What is required to receive the IR signal correctly is a program where the duration of the serial data bits can be measured and then the binary signal can be formed. Fortunately, the Arduino IDE provides a function called **pulseIn()** that can be used to measure the duration of a signal on a GPIO port. The format of this function is:

pulseIn(pin, value, timeout)

where, **pin** is the GPIO pin where the IR receiver output is connected to, **value** is HIGH or LOW depending on whether you want to measure the durations of HIGH or LOW signals respectively, and **timeout** (optional) specifies the maximum expected pulse duration in microseconds. The function returns the pulse duration in microseconds as an **unsigned long**, or a zero is returned if a timeout occurs.

As an example, the statement **dur = pulseIn(5, HIGH)** returns the duration of a pulse (HIGH) on pin 5 in microseconds. You will be using this function in your infrared control projects in the next sections to decode the data received by the IR receiver diode.

19.5 Project 1: Decoding the IR remote control codes

Description: In this project, you will use the supplied IR receiver diode and remote control unit to decode the signals sent when a button is pressed on the remote control unit. The received data will be displayed on the Serial Monitor.

Block diagram: Figure 19.5 shows the block diagram of the project.

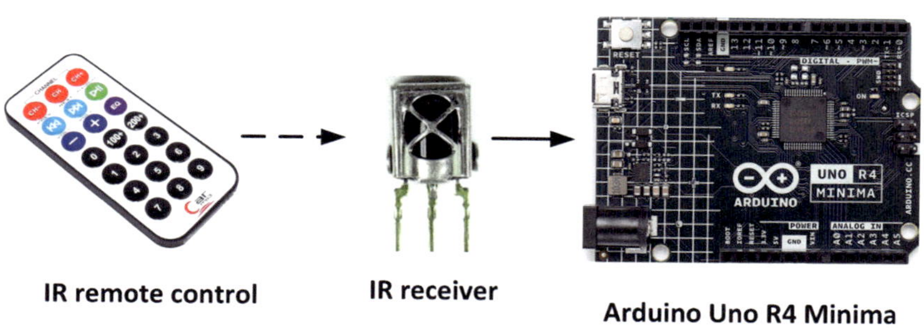

Figure 19.5: Block diagram of the project.

Circuit diagram: The circuit diagram of the project is very simple and is shown in Figure 19.6. The output of the IR receiver diode is connected to GPIO 2 of the Arduino Uno R4 Minima. Make sure that the power and ground pins are connected correctly.

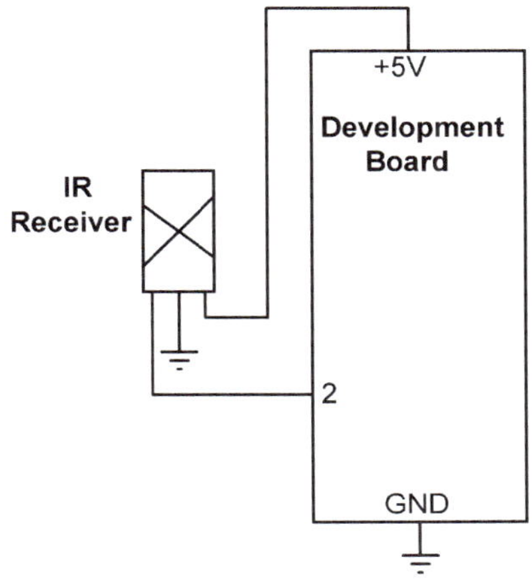

Figure 19.6: Circuit diagram of the project.

Program listing: Since there were no infrared libraries compatible with the Arduino Uno R4 at the time of authoring this book, the IR decoding was done using the first principles and referring to Figure 19.4. Figure 19.7 shows the program listing (Program: **IRTest**). At the beginning of the program, GPIO 2 is defined as the receiver port and is configured as a digital input. Function **IRreceive()** is where the IR signal is received and decoded. A **for** loop is formed to read 32 pulses. Inside this loop, if the pulse duration is greater than 1000 microseconds, then a 1 is assumed. Bitwise logical OR is used to set a bit after shifting left. The result is returned in variable code which is displayed on the Serial Monitor.

```
//---------------------------------------------------------------------
//                  INFRARED TRANSMITTER-RECEIVER TEST
//                  ==================================
//
// In this program the supplied IR receiver and remote control
// transmitter unit are used. The codes sent by the remote control
// when its buttons are pressed are displayed on Serial Monitor
//
// Author: Dogan Ibrahim
// File  : IRTest
// Date  : July, 2023
//---------------------------------------------------------------------
#define IRrx 2                                          // IR rX pin
unsigned long int shft, code = 0;

void setup()
{
```

```
  Serial.begin(9600);
  pinMode(IRrx, INPUT);                              // IR rx is input
}

unsigned long int IRreceive()
{
  while(digitalRead(IRrx) != LOW);                   // Wait for start (LOW)
  if(pulseIn(IRrx, HIGH, 15000) == 0) return 0;      // No start detected

  code = 0;
  for(int i = 0; i < 32; i++)                        // Receive 32 pulses
  {
    int pulseDuration = pulseIn(IRrx, HIGH, 3000);   // Timeout = 3 s
    if(pulseDuration == 0) return 0;                 // Timeout, exit
    if(pulseDuration > 1000)                         // 1 detected
    {
      shft = 1;
      shft = shft << i;                              // Shift left
      code = code | shft;
    }
  }

  while(digitalRead(IRrx) != HIGH);                  // Stop
  return code;
}

void loop()
{
  unsigned long res = IRreceive();
  if(res != 0) Serial.println(res, HEX);             // Display result in hex
}
```

Figure 19.7: Program: **IRTest**.

Test Procedure:

- Make sure that the remote control unit is loaded with a CR2025-type coin battery.
- Start the IDE, compile and upload the program.
- Start the Serial Monitor.
- Point the remote control unit to the IR receiver.
- Make a table of the received codes against the buttons pressed on the remote control unit. Table 19.1 was obtained by the author by pressing some of the buttons.

Button Pressed	Received Code
0	0xE916FF00
1	0xF30CFF00
2	0xE718FF00
3	0xA15EFF00
4	0xF708FF00
5	0xE31CFF00
6	0xA55AFF00
7	0xBD42FF00
8	0xAD52FF00
9	0xB54AFF00
+	0xEA15FF00

Table 19.1: Buttons pressed and received codes.

Figure 19.8 shows some of the codes displayed on the Serial Monitor.

```
E916FF00
F30CFF00
E718FF00
A15EFF00
F708FF00
E31CFF00
A55AFF00
BD42FF00
AD52FF00
B54AFF00
EA15FF00
```

Figure 19.8: Some of the codes displayed.

Now that you have the IR codes of your remote control unit, you can develop many interesting IR remote control-based projects. A simple project is given in the next section.

19.6 Project 2: Remote relay activation/deactivation

Description: In this project, a relay is added to the project. The operation of the project is as follows: When button 0 is pressed on the remote control, the relay turns ON. Similarly, when button 1 is pressed, the relay turns OFF. This way, you can control any device or appliance remotely by sending codes from the remote control unit.

Block diagram: Figure 19.9 shows the bloc diagram of the project.

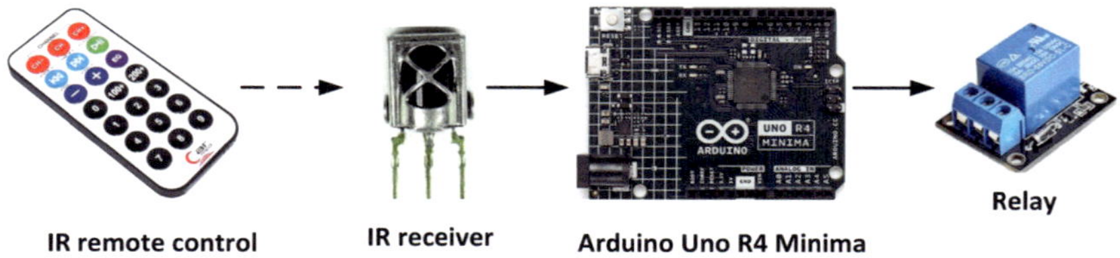

Figure 19.9: Block diagram of the project.

Circuit diagram: The relay is connected to port 3 of the development board as shown in Figure 19.10

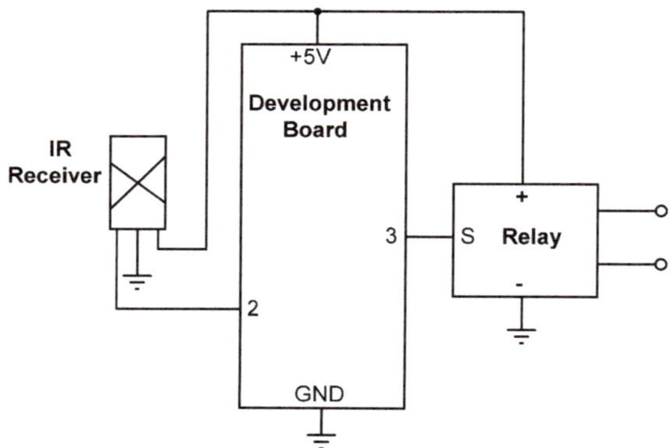

Figure 19.10: Circuit diagram of the project.

Program listing: Figure 19.11 shows the program listing (Program: **REMOTECTRL**). At the beginning of the program, the IR receiver is assigned to port 2 and **RELAY** is assigned to port 3 and the ON and OFF codes are defined as shown in Table 19.1. Inside the **setup()** function, **RELAY** is configured as output and it is deactivated. Also, the on-board LED (at port 13) is configured as output and turned OFF. Inside the main program loop, the program waits to receive code. When button 0 is pressed, the relay is turned ON. Similarly, when button 1 is pressed, the relay is turned OFF. The LED blinks as soon as a code is received from the remote controller.

```
//-----------------------------------------------------------------------
//                INFRARED REMOTE CONTROL OF A RELAY
//                ==================================
//
// In this program the supplied IR receiver and remote control
// transmitter unit are used. A relay is connected to the development
// board. Pressing button 0 activates the relay. Pressing button 1
// deactivates the relay. On-board LED blinks to indicate button press
//
// Author: Dogan Ibrahim
// File  : REMOTECTRL
// Date  : July, 2023
```

```
//-----------------------------------------------------------------------
#define IRrx 2                                          // IR rX pin
unsigned long int shft, code = 0;

int RELAY = 3;                                          // RELAY at port 3
int LED = 13;                                           // On-board LED
#define ON 0xE916FF00                                   // RELAY ON code (0)
#define OFF 0xF30CFF00                                  // RELAY OFF coe (1)

void setup()
{
  pinMode(IRrx, INPUT);                                 // IR rx is input
  pinMode(RELAY, OUTPUT);                               // RELAY is output
  digitalWrite(RELAY, LOW);                             // Deactivate relay
  pinMode(LED, OUTPUT);                                 // LED is output
  digitalWrite(LED, LOW);                               // LED OFF
}

unsigned long int IRreceive()
{
  while(digitalRead(IRrx) != LOW);                      // Wait for start (LOW)
  if(pulseIn(IRrx, HIGH, 15000) == 0) return 0;         // No start detected

  code = 0;
  for(int i = 0; i < 32; i++)                           // Receive 32 pulses
  {
    int pulseDuration = pulseIn(IRrx, HIGH, 3000);      // Timeout = 3 s
    if(pulseDuration == 0) return 0;                    // Timeout, exit
    if(pulseDuration > 1000)                            // 1 detected
    {
      shft = 1;
      shft = shft << i;                                 // Shift left
      code = code | shft;
    }
  }

  while(digitalRead(IRrx) != HIGH);                     // Stop
  return code;
}

void loop()
{
  unsigned long res = IRreceive();
  if(res != 0)
  {
    if(res == ON) digitalWrite(RELAY, HIGH);            // RELAY ON
```

```
        else if(res == OFF)digitalWrite(RELAY, LOW);        // RELAY OFF

        digitalWrite(LED, HIGH);                           // Flash LED
        delay(500);
        digitalWrite(LED, LOW);
    }
}
```

Figure 19.11: Program: **REMOTECTRL**.

19.7 Project 3: Infrared remote stepper motor control

Description: In this project, the number of revolutions of a stepper motor (28BYJ-48) is controlled using the IR receiver and remote control unit. For example, pressing 5 rotates the motor by 5 full revolutions. Pressing 12 rotates the motor by 12 full revolutions, etc. Numbers are entered from the remote control unit and must be terminated with the **+** button. For example, to rotate 5 times enter **5+**, similarly, to rotate 15 times, press **15+**. It is assumed that only clockwise rotation is required (anticlockwise rotation can easily be added if required).

Block diagram: Figure 19.12 shows the block diagram of the project.

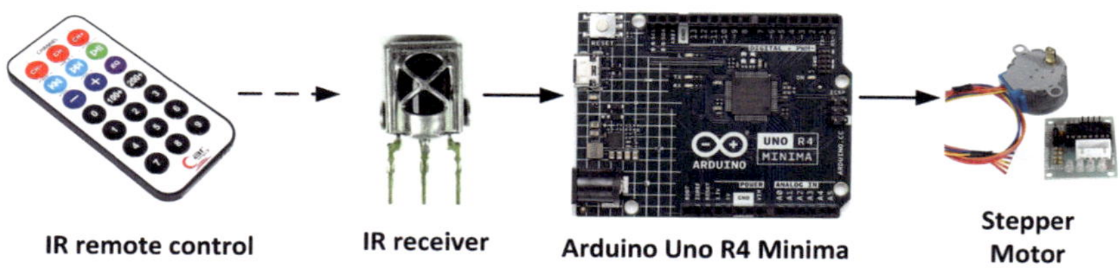

Figure 19.12: Block diagram of the project.

Circuit diagram: The stepper motor is connected to port pins 8, 9, 10, and 11. The IR receiver is connected to port 2 as in the previous project. Figure 19.13 shows the circuit diagram.

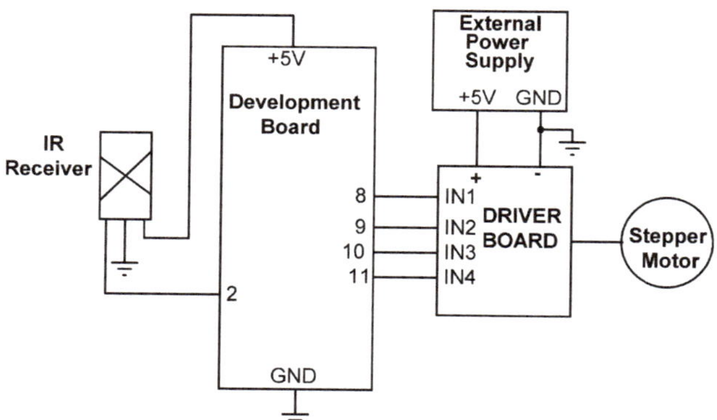

Figure 19.13: Circuit diagram of the project.

Program listing: Figure 19.14 shows the program listing (Program: **IRSTEPPER**). At the beginning of the program, the remote control, and then the stepper motor details are given. The IR receiver is connected to port 2. Stepper motor connections must be in the form IN1-IN3-IN2-IN4 as shown in the program. Then the remote control codes are defined for all the numbers 0 to 9 and for the **+** sign which is used as the terminator. Motor steps per revolution is set to 2048 (it is actually 2038 but 2048 seems to give more accurate results). Inside the **setup()** function, the motor speed is set to 9 RPM (setting to higher or lower values may give incorrect rotations). Function **GetNumber()** receives the IR code as its argument and returns the numeric value of the button pressed on the remote control unit. Inside the main program loop, the text **Ready...** is displayed when the program is ready to read a number. The number entered by the user is displayed. When the **+** terminator button is pressed, the required number of revolutions is displayed and the motor starts rotating. The total number of steps sent to the motor is calculated by multiplying the required number of revolutions (**Sum**) by the **StpsPerRev**.

```
//----------------------------------------------------------------------
//                  REMOTE IR STEPPER MOTOR CONTROL
//                  ===============================
//
// In this program the supplied IR receiver and remote control
// transmitter unit are used to control the number of full rotations
// of the stepper motor
//
// Author: Dogan Ibrahim
// File  : IRSTEPPER
// Date  : July, 2023
//----------------------------------------------------------------------
#include <Stepper.h>
const int StpsPerRev = 2048;                            // Steps per rev
Stepper MyStepper = Stepper(StpsPerRev,8,10,9,11);

#define IRrx 2                                          // IR rX pin
unsigned long int shft, code = 0;

//
// IR remote control codes
//
#define N0 0xE916FF00                                   // Code 0
#define N1 0xF30CFF00                                   // Code 1
#define N2 0xE718FF00                                   // Code 2
#define N3 0xA15EFF00                                   // Code 3
#define N4 0xF708FF00                                   // Code 4
#define N5 0xE31CFF00                                   // Code 5
#define N6 0xA55AFF00                                   // Code 6
#define N7 0xBD42FF00                                   // Code 7
#define N8 0xAD52FF00                                   // Code 8
```

```
#define N9 0xB54AFF00                                          // Code 9
#define PLUS 0xEA15FF00                                        // Code +

void setup()
{
  pinMode(IRrx, INPUT);                                        // IR rx is input
  MyStepper.setSpeed(9);                                       // Speed=9 RPM
  Serial.begin(9600);
}

//
// This function returns the numeric value of the pressed button
//
int GetNumber(int m)
{
  switch(m)
  {
    case N0:                                    // 0 pressed
      return 0;
      break;
    case N1:                                    // 1 pressed
      return 1;
      break;
    case N2:                                    // 2 pressed
      return 2;
      break;
    case N3:                                    // 3 pressed
      return 3;
      break;
    case N4:                                    // 4 pressed
      return 4;
      break;
    case N5:                                    // 5 pressed
      return 5;
      break;
    case N6:                                    // 6 pressed
      return 6;
      break;
    case N7:                                    // 7 pressed
      return 7;
      break;
    case N8:                                    // 8 pressed
      return 8;
      break;
    case N9:                                    // 9 pressed
      return 9;
```

```
      break;
    case PLUS:                              // + pressed
      return -1;
      break;
  }
}

//
// Receive a code and decode it
//
unsigned long int IRreceive()
{
  while(digitalRead(IRrx) != LOW);                    // Wait for start (LOW)
  if(pulseIn(IRrx, HIGH, 15000) == 0) return 0;       // No start detected

  code = 0;
  for(int i = 0; i < 32; i++)                         // Receive 32 pulses
  {
    int pulseDuration = pulseIn(IRrx, HIGH, 3000);    // Timeout = 3 s
    if(pulseDuration == 0) return 0;                  // Timeout, exit
    if(pulseDuration > 1000)                          // 1 detected
    {
      shft = 1;
      shft = shft << i;                               // Shift left
      code = code | shft;
    }
  }

  while(digitalRead(IRrx) != HIGH);                   // Stop
  return code;
}

void loop()
{
  int Sum = 0;
  Serial.println("Ready...");
  while(1)
  {
    unsigned long res = IRreceive();                  // Receive code
    if(res != 0)
    {
      int N = GetNumber(res);                         // Get a number
      Serial.println(N);
      if(N == -1)break;                               // If terminated
      Sum = Sum * 10 + N;                             // Required number
    }
```

```
  }
  Serial.print("Number of revolutions: ");
  Serial.println(Sum);
  MyStepper.step(Sum * StpsPerRev);
  delay(100);
}
```

Figure 19.14: Program: **IRSTEPPER**.

Index

A
Accurate clock 149
analogWave 242
Animation 267
Arduino Web Editor 23
ATmega328P 14
Attachinterrupt 82

B
Binary counting 72
Blanking leading zeroes 124
Blinking LED 59
Bluetooth 276
Bluetooth BLE 277
Boards manager 25
Builtin RTC 216
Buzzer 174

C
CAN bus 292
CAN bus node 295
CAN bus arbitration 297
CAN bus transceivers 300
CAN connectors 298
Chaser LEDs 67
Cortex-M4 15
Creating images 262
Current sinking 61
Current sourcing 61
Custom LCD characters 144

D
DAC 241
Darkness reminder 164
DHT11 181
Displaying shapes 227
Door security lock 203
DS1302 207

E
EEPROM 248
Espressif S3 15
External interrupt 80

F
Factorial 54
Flame sensor 177

H
HID 14
Humidity sensor 180

I
I^2C bus 136
I^2C LCD 137
I^2C ports 137
Infrared controller 311
Infrared receiver 311
Integer calculator 198

J
Joystick 221

K
Keypad 194

L
LCD 136
LCD dice 154
LDR 146
LED dimming 255
LED matrix 226, 260
LM35 157

M
Matrices 56
Melody maker 184
MFRC522 189
Multiplexed LED 114

P
Periodic interrupt 218
Println 27
PWM 248

Q
Quadratic equation 50

R
RA4M1 18
Random flashing 74, 103
Reaction timer 83, 128
Renesas RA4M1 14
RFID reader 187
RTC 207

S
Scrolling text 142
Serial communication 280
Serial monitor 26
Servo motor 231
Seven segment LED 109
Shift register 100
Simulator 285
Sine wave 243
Sketch 23
Sound detection 175
Square wave 241
Stepper motor 237

T
Tag ID 187
TCP 269
Temperature controller 161
Tilt detection 167
Timer interrupt 119
Time stamping 213
Traffic lights 89, 94

U
UDP 269
USB-C 14

V
Vibration tilt sensor 167
Voltmeter 160

W
Watchdog timer 120
Water level controller 172
Water level sensor 169
WiFi 257
Wiring 12
WokWi 286